Encouraging Change

Sustainable Steps in Water Supply, Sanitation and Hygiene

Sally Sutton and Hope Nkoloma

With contributions from :

Fordson Nyirenda

Isaac Mbewe

John Kelleher

The original English version of this manual, including the research on which it was based, was partly funded by the following organisations:

St. Albans, October 2011

Current edition: October 2011

First published April 2003 by TALC now Health Books International

Reprinted in 2020
by Practical Action Publishing
Rugby, Warwickshire UK

Editor: TALC – Teaching Aids at Low Cost
Authors: Sally Sutton and Hope Nkoloma
Layout: Isabelle Ciret and Alison Jones
Cover Design: Jamie Thompson
Cover Photos: E. Kelly, S. Sutton and V.Curtis
Printing: Short Run Press
ISBN: 978-0-9558811-8-3

Practical Action Publishing Ltd
27a Albert Street, Rugby, CV21 2SG, Warwickshire, UK
www.practicalactionpublishing.org

Also available in CD-ROM and digital formats form.
www.healthbooksinternational.org

This publication has been produced with support from The Trusthouse Charitable Foundation.

The Trusthouse
Charitable Foundation

TABLE OF CONTENTS

POSTERS

Posters are in thematic series, but each is designed to be used for several different modules/ purposes. They are shown in miniature in the text but are also available as A4 posters at the back of this manual. Those included so far are -:

G= General themes
G 1–19 Daily activities / times of the day / stages of house building
(See pages 41-42, 48, 54)

G.1 Dawn/getting up	G.11 Bathing
G.2 Eating a meal	G.12 Cleaning up the baby
G.3 Cooking/preparing food	G.13 Boiling water
G.4 Working in the fields	G.14 Going to the clinic
G.5 Cleaning around the house	G.15 Fields
G.6 Drinking beer	G.16 Building a house
G.7 At night	G.17 Large house
G.8 Washing clothes	G.18 Small house
G.9 Washing dishes	G.19 Costs of sickness
G.10 Collecting firewood	

H= Hygiene themes
H1-14 Hygiene practices for Faecal Transmission diagram and 3 pile sorting (See page 22)

H.1 Hand	H.8 Untidy house
H.2 Mouth	H.9 Throwing rubbish into a pit
H.3 Baby's face with flies	H.10 Throwing rubbish onto a heap
H.4 Dog licking the plates	H.11 Dipping water container into a well
H.5 Uncovered food with flies	H.12 Washing children's hands
H.6 Covered food	H.13 Pot rack for drying dishes
H.7 Foot	H.14 Faecal-oral Transmission ('F') Diagram

H21-34 Handwashing practices (See pages 38, 39, 91–100)

H.21 Into bowl, without soap	H.28 Clean hands -social benefits of handwashing
H.22 Into bowl, with soap	H.29 Making a tippy-tap (1)
H.23 Scooping out of bowl, no soap	H.30 Making a tippy-tap (2)
H.24 Scooping out of bowl, using soap	H.31 Calabash handwasher
H.25 Poured from jug, no soap	H.32 Leaking ladle handwasher
H.26 Poured from jug using soap	H.33 Handwashing Tips
H.27 Motivation by peers	H.34 Constructions for handwashing

S= Sanitation themes

S1-10   Types of latrine (See pages 56, 81)

S.1 Traditional latrine (basic, unimproved)	S.6 Improved pit latrine (2)
S.2 Traditional latrine, improved (grass-roofed)	S.7 Brick/mud-brick improved pit latrine, VIP
S.3 Traditional latrine improved (VIP)	S.8 Improved pit latrine (3) with square Sanplat
S.4 Traditional improved latrine	S.9 Ventilated Improved Pit Latrine (VIP) – how it works
S.5 Ordinary pit latrine (1)	S.10 Ecological Sanitation / Composting / Urine Diverting latrine

S11- 17 Defecation practices (See page 82)

S.11 Going in the bush	S.15 Going in the field
S.12 Defecating in a lake, dam or river	S.16 Burying faeces with a hoe (Cat Method 1)
S.13 Using a stream	S.17 Mother cleans up after child (Cat Method 2)
S.14 Open defecation near houses	

S20-22   Types of squat hole (See page 81)

S.20 SanPlat	S.22 Wooden squat hole
S.21 Squat hole lined with mortar or cement	

W= water supply

W1-19   Main (mostly protected) source types, including water lifting devices (See pages 56, 65-66 and 75)

W.1 Handpump on borehole	W.11 Unimproved spring
W.2 Handpump on shallow well	W.12 Improved spring
W.3 Bucket pump on borehole / hand-augered well	W.13 Unimproved scoophole
W.4 Lined well with windlass	W.14 Roof catchment of rainwater
W.5 Improved family well (1)	W.15 Rainwater catchment (especially Schools and Rural Health Centres)
W.6 Improved family well (2)	W.16 Bucket and beam lifting device
W.7 Improved family well (3)	W.17 Rope pump
W.8 Unimproved family well	W.18 Treadle pump
W.9 Improved scoophole (1)	W.19 Unimproved surface water source
W.10 Improved scoophole (2)	

W20-28 Methods of transporting water (see Page 88)

W.20 On the head and by hand (1)	W.25 On the head and by hand (2)
W.21 On the shoulders	W.26 Using oxen
W.22 By hand	W.27 Using a bicycle
W.23 All the family carrying water on head and by hand	W.28 Carrying containers
W.24 By wheelbarrow	

W29-36 Options for lining shafts (wells and latrines) (see Page 74)

W.29 Shallow brick-lining or masonry for spring, scoophole or latrine pit	W 33 Partial concrete ring lining
W.30 Fully brick- lined well with concrete rings below highest water level	W.34 Fully lined with concrete rings
W.31 Fully lined well of brick and stone, with 'dry stone' below highest water level	W.35 Partial wood lining
W.32 Partial brick lining, old and new	W.36 Basket weave lining for pits

W37-47 Minimum options for the top of water sources, step by step improvement (see Page 75)

W.37 Unprotected well or scoophole	W.43 Pole to hang bucket and rope
W.38 Mouth of shaft built up (mound) to stop flow into source	W.44 Locally made windlass
W.39 Source closed with oil drum and lid	W.45 How a pulley works
W.40 Top strengthened with brick top wall and small apron	W.46 Using a pulley to draw water from a well (1)
W.41 Source with apron and drainage (soak-away)	W.47 Using a pulley to draw water from a well (2)
W.42 Source with apron, drainage and garden	

NB All posters included in this document are available as A4 posters at the back of the manual.

Poster drawings are by the following

Björn Brandberg

Lois Carter

The CEP group, Mongu, Western province, Zambia

Jane Lingard

LINTAS

Peter Morgan

Odia

Sally Sutton

Jamie Thompson

WATERAID

All photographs are by Sally Sutton, unless indicated otherwise.

LIST OF TABLES

LIST OF FIGURES

FOREWORD

Every 20 seconds a child dies from diarrhoeal disease caused by a lack of clean water, hygiene and sanitation. In fact, by the time that you finish reading this foreword, it is likely that another five children will have died. This equates to 4,000 children per day or 160 primary school classrooms of children. Shocking though these statistics undoubtedly are, they represent only a fraction of the suffering caused by water-related illnesses, much of which cannot be counted or measured. There are the lost wages caused by absence from work, the impaired brain function due to parasitic worms, the hours spent caring for a sick child that could be passed in school or undertaking vital household tasks.... To those living or working in communities where access to clean water is limited or non-existent, this endless list of problems is all too familiar.

Of course the great tragedy is that so much of this is preventable. When we consider the undoubtedly vast scale of suffering, it seems almost incredible that something as simple as dirt on an unwashed hand could be responsible, yet time and again it has been demonstrated that even elementary changes in behaviour such as hand-washing and correct disposal of faeces can have a dramatic impact upon the health and well-being of communities.

Encouraging Change sets out to implement these changes, providing simple but effective solutions to improving water quality and hygiene. Developed in Zambia to meet the needs of rural communities with minimal funds, the manual guides communities and health workers step-by-step through the process of change: identifying problems, planning and implementing change, and finally evaluating the success of these new developments.

The book takes as its starting point the concept that change can only be truly effective with the support of communities themselves and focuses on empowering villagers to embrace and direct the process of change. Throughout the process of change, the role of the health worker is that of facilitator, offering essential guidance and support, whilst ensuring that the natural leadership of any project remains with the community.

The processes in the manual ensure that each individual, regardless of age or gender, can make a valuable contribution to the health of their community. Activities and discussions are designed specially to ensure that groups who might traditionally have been excluded from decisions are able to express their views freely and can take an active role in both implementing and maintaining changes.

Sadly, we cannot simply wave a magic wand and make the world's water instantly safe and clean tomorrow. However, as the success of *Encouraging Change* in Zambian communities has demonstrated, the united efforts of communities and the health workers supporting them can achieve genuine, lasting change.

Sandy Cairncross
Professor of Environmental Health
London School of Hygiene and Tropical Medicine
September 2011

ACKNOWLEDGEMENTS

We would firstly like to thank Lois Carter and Alison Jones (TALC) for managing the transference of this manual into a form suitable for use in different countries. We would also like to thank Professor Sandy Cairncross for his helpful advice.

We would like to thank all those in Zambia who helped in the development of this manual. In particular, Fordson Nyirenda (Environmental Health Specialist CBOH – Central Board of Health), Isaac Mbewe (Head of NWASHE and then with WaterAid) along with John Kelleher of WaterAid, provided many good ideas and much editorial input. Adam Hussen (Director of Water Affairs) and Clement Mwandwe (Public Health, Food and Drugs Laboratory) provided considerable support and promotion of the ideas, without which the concept would not have taken root.

Grateful mention should also be made to the Environmental Health Technicians in Northern, Western and Luapula Provinces who carried out much of the fieldwork and to Erik Kelly of Peace Corps who helped in training and workshops.

Management and co-ordination of activities was initially carried out by Isaac Mbewe of NWASHE, and later by Kenneth Nyundu (DWA - Department of Water Affairs) assisted at provincial levels by Osward Mwansa, Daniel Mwanza (DWA Kasama) and Stephen Filumba (DHMT Kasama) George Serenje (DHMT Mansa) and the DHMT of Kaoma (especially Messrs Mwanamwambo, Kalombo, Nkhata) and the ladies of CEP Mongu (Catherine Akekelwa, Albertina Siyunyi, Georgina Akolwa, Precious Muuka, Eva Mbuuwa and Clare Mulopo). CEP were also involved in developing some of the modules and illustrations, and owe much to Joanne Harnmeijer for the original creation of the group.

When we were uncertain whether the manual was veering off course, we received encouragement and comments from Sarah House, Joy Morgan, Vicky Blagborough and Kathy Attawell.

Mention should also be made of those who played a part in the research who sadly passed away before its completion. They include -:

M. Samani NWASHE Rural water supply engineer
J. Mathe NWASHE Health Educator
Mr Tshintu MLGH (Ministry of Local Government and Housing)
L. Banda Laboratory technician Mongu
M Simakando DHI (District Health Inspector) Mongu

Sally Sutton, Hope Nkoloma.   September 2011

GLOSSARY

Aquifer	Underground layer of soil, sand or rock that yields water
Aeolian sands	Wind-blown sands
Alluvial sands	Sands transported or laid down by water
Apron	Concrete area surrounding the top of the well, intended to catch any spilt water and direct it away from the well and into the drainage channel/soakaway.
Bailing	Removing water from a well by repeatedly filling and emptying a bucket. This is usually done when it is necessary to clean the well. The bucketfuls of water are called ***bailings.***
Baseline data	Information that is gathered about a situation before a change or project is begun. This can be compared to data gathered after the project has been completed to measure the impact of the changes.
Borehole	A narrow wellshaft (less than 0.5 metres diameter) created by drilling with a mechanical rig or manually.
Casing	A large, smooth pipe lowered down a borehole and left there, to prevent the sides from falling in.
Chlorine	Chemical used to disinfect and clean wells.
Cistern	A container or tank used for storing water.
Community Sanitation Strategy	A plan for improving the sanitation within an area
Consolidated ground	Ground that stands up easily and is hard to dig. It is often a good location for wells, as it is less likely to collapse. ***Unconsolidated ground*** collapses easily and is not cemented.
Coverage	The proportion of the population who have access to a given service
Credit System	An arrangement through which people can borrow money
Cylinder Well	Hand-drilled, shallow borehole
Defecation	Depositing faeces
Diameter	The width across the widest part of a circle
Drainage channel	A small trench diverting spilt water or rainwater away from the well. Water flows from the apron into the drainage channel.
Ecological sanitation system / Composting latrine	A kind of toilet which stores faeces in a container in which they are transformed into compost.
Extension worker	Field or sub-district level worker
Facilitator	The person in charge of a lesson or workshop

Faecal coliform count (FC count)	A measurement of water quality that demonstrates how far water has been contaminated by faeces. The lower the number, the safer the water. Ideally, water should have zero faecal coliforms/100ml.
Faecal-oral transmission routes	The ways in which faeces can accidentally get into the mouth, causing diarrhoeal disease.
Filtration	A process of water treatment in which the water is passed through a medium, usually a lot of sand
Freshwater lens	A layer of fresh water in an aquifer, which is limited in extent by underlying salt water..
Groundwater	Water that has soaked deep into the ground.
Hand- augering	Creation of a hole/borehole in the ground using a type of hand tool known as an auger.
Headworks	Structures built on top of a water source e.g. parapet and apron, spring catchment box, pump, etc.
Irrigation	Watering soil, usually for the purpose of growing crops.
Jetting	Creation of a borehole in the ground using a high speed jet of water to dig.
Lining	The materials used on the inside of the well shaft e.g. concrete rings/blocks, bricks, stones, wood. Lining reduces the risk of the well collapsing and prevents dirt from getting inside. When it is only used on part of the shaft it is known as *Partial Lining* (see Posters W34, W35).
Matrix	A chart that works in several directions (see example, p.31)
NGO (Non-Governmental Organization)	A private organisation involved in community development.
Operations and Maintenance	Actions taken to keep the facility working.
Oral Rehydration Solution (ORS)	A mixture of dry salts and water that is given to people with diarrhoea to prevent them becoming dehydrated.
Parapet / Surround	The built-up, wall-like structure around the well mouth.
Permeability	How easily rock / soil / sand allows water to pass through it
Pilot projects	Projects implemented initially in a small local area to test the practicality of their design – and then to demonstrate it.
Ponding	Build-up of surface water in an area.
Pulley	A small wheel mounted on a bar over the well mouth, to make it easier to lift buckets of water from the bottom of the well
Protected sources	Sources which have been built on to protect the water from contamination e.g. a spring box, a well cover.
Rebar	Steel bar used for reinforcing structures, preventing them from collapsing

Rope pump (low-cost)	Simple pump often using locally available materials. It lifts the water with a rope and disks (car tyre or plastic) inside a small diameter pipe. (See Poster W17)
Run-off	Water from rain that runs away along the surface of the ground
Saline	Salty
Schistosomiasis (also known as Bilharzia)	A disease caused by parasites that live in snails in stagnant water.
Scoophole	A water hole dug by hand to obtain shallow groundwater through scooping it out by hand.
Sanplat	A concrete squatting slab that can be placed over a traditional latrine to improve hygiene. It is easy to wash and reduces flies and smells. (See Poster S20)
Spring / Seepage point	A point where water trickles out of the ground, creating a pool or stream. They are found where the aquifer meets the surface of the ground.
Soakaway	A pit containing rocks where dirty water leaving the drainage channel can soak into the ground. In some instances a garden can be used instead.
Source of water	A place where water can be obtained.
Top slab	Concrete slab with a hole in the centre, situated on top of the well mouth. Together with the *Cover* it prevents dirt, animals or people from falling into the well.
Treadle pump	Water is lifted by suction from as deep as 7 metres below ground level. It is a system especially suitable for irrigation (see Poster W18)
Vent pipe	Chimney-like pipe protruding above the roof of a latrine. Increases the air moving out of the latrine, which reduces smells. If it is closed at the top with mosquito mesh, it will also prevent flies from breeding in the pit.
Water-flushed system	A toilet with a water seal which is flushed after use.
Water-logging	The ground becomes saturated with water. Any excess water will sit on top of the ground rather than sinking into it.
Water seal	U-bend full of water to prevent smells, insects etc. coming into the toilet from the pit or sewer.
Water table	The level below which the ground is saturated with water.
Weathered layer	A layer of rock/soil that has been worn away or broken down by exposure to the weather.
Well head	All the parts of a well above ground level. (This ideally includes top slab, cover, lining, apron, drainage channel / soakaway and either a pole or lifting device.)
Well shaft	The part of the well from ground level to the bottom of the well.
Well mouth	The opening at the top of a well shaft.
Windlass	A device used for lifting water out of a well (see Poster W47)

METHODS

Role-play	Participants act out a situation or story.
Focus/Focal Group Discussion	The group discusses a topic.
3-pile sorting	Participants are given posters of different activities. They decide if these are 'Good Practice', 'Bad Practice' or 'Neither Good Nor Bad Practice' and sort them into three piles.
Story with a gap	Participants are given two pictures or descriptions of a situation 'before' and 'after' change. They are asked to describe what happened in between.
Seasonal / Seasonality Calendar	A list of what people do (or intend to do) in different months of the year.
Pocket Chart Voting	Posters are arranged on the ground in a chart or matrix. Participants put counters or stones on the chart to show which of the activities they do, and when.
Body Mapping	Participants draw the outline of a body and then discuss how different living conditions can affect the different body parts.
Pair-wise ranking	Participants take two pictures and decide which is more important. They then compare the 'more important' picture to another picture and do the same again. This is repeated until all pictures are in order. Equal ranking can be used if necessary.
Semi-structured interview	One–to-one information collection with a check-list of topics to be covered and key questions to ask.
History line / Historical timeline	Participants create a line that represents the history of their community and discuss the key events and changes that should be marked on it.
Blind voting	Participants give their answer to a question secretly so that the rest of the group do not know how they answered. Usually done by asking people to place counters on a chart or matrix.
Sanitation ladder	Participants are given posters showing different sanitation options. They place these in a 'ladder' according to their preferences.
Open-ended story	Facilitator begins a story and asks the participants to finish it, discussing the key issues.
Transect walk	Facilitator walks around the village, noting down any useful information about the community and people.
Project Cycle	The different activities and stages involved in a project, arranged in the order in which they happen.

The parts of a well

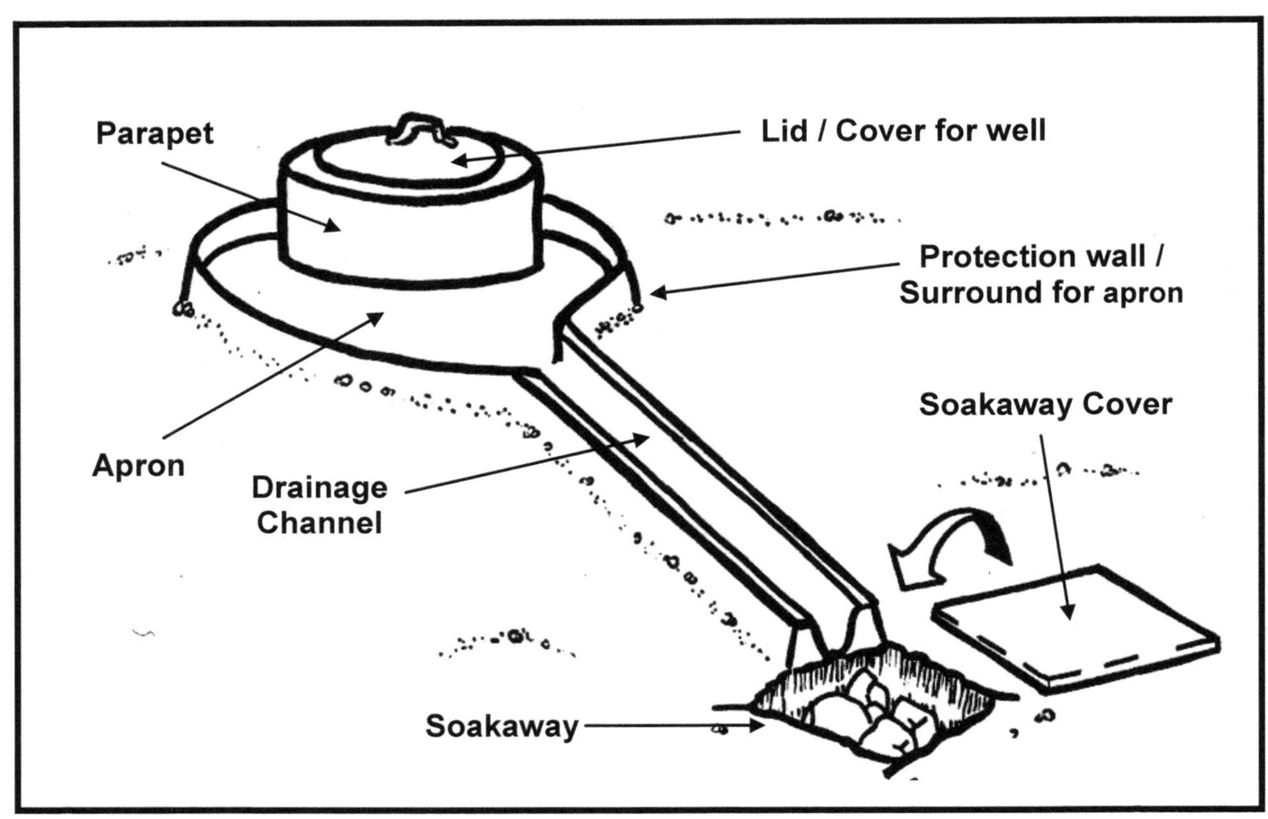

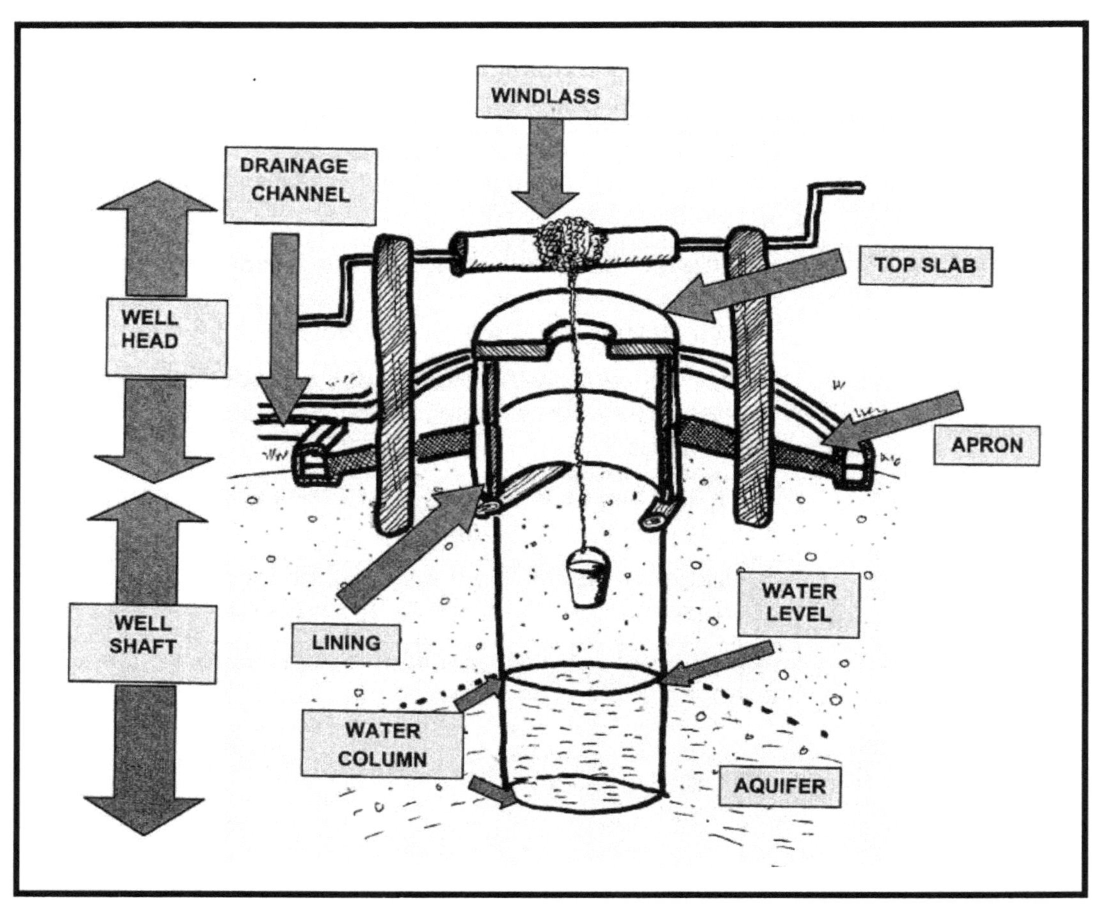

SECTION ZERO

GETTING TO KNOW THE COMMUNITY AND IDENTIFYING KEY ISSUES

0.1 CONTEXT OF THE MANUAL

0.2 INTRODUCING CHANGE

0.3 NOTES TO FACILITATOR

0. BACKGROUND INFORMATION

0.1 The context of the manual

This manual on **'Encouraging C hange'** is a field-level facilitator's guide which can be supplemented by technical guidelines explaining the practicalities of how water supplies and sanitary facilities can be constructed or improved (see References page 126). It has been developed in Zambia and is being used by the Ministry of Health as a reference document for extension workers, and as the basis for training in participatory methods. The idea is that it provides a framework in a form that can be 'customised' by people in different countries so that the facts, examples and posters can reflect the local situation.

The methods described can equally well be used for other community and health issues, such as malnutrition and childcare, malaria and bilharzia prevention, or establishing credit systems or co-operatives. The methods are very flexible, and you are invited to send in examples of how else you have found they can be used. This manual is just a start.... We hope that you will help improve it and make it more useful to people working at grass-roots level and those planning how to introduce change. Your comments will be therefore be very useful and you can find a note on how to give us feedback at the back of this manual.

0.2 Introducing change

This volume provides some ideas and materials to help bring about **improvement in water, sanitation and hygiene**, even where outside assistance is not available. It provides education materials to help communities decide on what changes to make, and how to plan and implement these changes. It also seeks to make them feel capable to undertake change and to look positively on their assets.

Whilst there will be occasions where the facilitator or the community may be able to call upon additional funds from outside, the emphasis is **on l ow co st or no cost changes.** There are considerable and effective changes that communities can make with little or no additional resources, if they firmly believe they can, and they plan effectively. However, where outside funding is available, the options are wider, but it may be more difficult for others to do the same. In both cases, keeping costs as low as possible will increase the number of people who can benefit and the long term sustainability.

The framework outlined in this manual reflects the **process** (see Steps Towards Change, Fig 0.1) by which this can be achieved, using as far as possible the ideas and skills available within the community. To promote discussion and involve all groups in the community, it uses participatory methods which encourage everyone to play an active part in the process of change. This approach has been shown to be more effective than simply telling people what to do, and helps avoid the situation where dominant people in the community decide on what changes to make while others who do not agree cannot express their thoughts and influence decisions.

Encouraging people to change their situation and/or behaviour involves building or reinforcing their belief in the need for change and in their capacity to bring about that change. Peer pressure and the establishment of community-led standards of behaviour are important elements in spreading and maintaining changes. Without these, the effects of any actions will tend to be very short-lived.

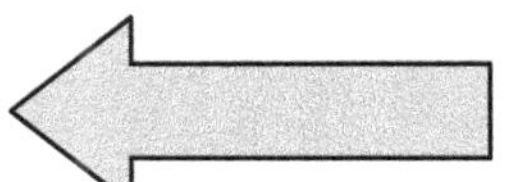

The process of change is often slow, requiring people to readjust their views, focus on key issues and work together as a team. The examples given in this manual illustrate what can be achieved. However to carry out all the modules included would impose an unacceptable workload where facilitators plan to work in several communities. The ideas is therefore more to pick out or modify ones which seem most appropriate to the situation in specific communities, and help in the design of meetings you are planning. The overall framework of moving from problem identification through analysis and solutions to planning, implementation and evaluation is an important one which will help communities to achieve changes they may not presently realise they can achieve.

Table 0.1 STAGES AND THE SAMPLE SESSIONS INCLUDED IN THE MANUAL

STAGES	Components (select relevant ones from examples given)	Methods (FGD = Focal Group Discussion)
ONE Getting to know the community and identifying issues causing most concern	**1.1** Getting to know the place. **1.2** Identification of key issues (general) **1.3** The costs of being sick **1.4** Identification of key (health) issues	Transect walk Community mapping Role play, FGD Blocking transmission routes/ barrier 'F' diagram.
TWO Analysing problems and their underlying causes	**2.1** What is wrong with our water supply? **2.2** Do people's defecation practices cause problems? **2.3** Can hands really spread disease? **2.4** Do our handwashing practices cause any problems? **2.5** Do we have enough water in the house?	Drawing , FGD Pocket chart voting, FGD Role play, FGD Pocket chart voting Daily activity chart
THREE Building community capacity to change the situation	**3.1** Building on experience **3.2** Project cycle **3.3** What resources and skills can be made available? **3.4** Increasing affordability with improved facilities **3.5** Doing things step by step	History timeline Story with a gap Mapping/seasonal calendar Story with a gap Story with a gap
FOUR Looking for solutions, selecting options.	**4.1** and **4.2** Identifying technology options for water supply **4.3** Selecting preferred options for sanitation. **4.4** and **4.5** Developing behavioural standards at community and household level in water collection and water use **4.6** Encouraging hand washing **4.7** Household hygiene **4.8** and **4.9** Personal and household hygiene	Ladder ranking/ matrix Ladder ranking/ matrix Role play/ daily activities FGD, posters FGD, Body mapping, 3 pile sort
FIVE Planning how changes will be made	**5.1** What are we going to do (community level) **5.2** Task allocation and task sharing (gender) **5.3** Planning for changes at household level **5.4** Planning for behavioural change at individual level	Seasonal calendar Daily activity chart FGD
SIX Implementation, Monitoring and Evaluation of Project	**6.1** Monitoring progress **6.2** Assessing Achievements **6.3** Evaluating Changes **6.4** Evaluating the Impact	Seasonal calendar + plan from 5, Drawing, body mapping, FGD

Note: You can use Table 0.1 as you go on, to help plan what you will do in each community. Section 3 Modules for community preparation may be used at any stage. They may sometimes be in demand after options have been explored, but tend to be more necessary as a capacity building exercise before these are considered.

If there are priority concerns from the start, they can be focused on by limiting the modules used, as follows -:

Table 0.2 Grouping of modules according to topic

PRIORITY CONCERNS	PROBLEM ANALYSIS	COMMUNITY PREPARATION	DEFINING OPTIONS	ACTION PLANNING
Faecal disposal	2.2	3.1-3.5	4.3	5.3+5.2
Safe water supply	2.2	3.1-3.5	4.1, 4.2, 4.4	5.1+5.2
Handwashing/ personal hygiene	2.3-2.5	3.2+3.3	4.5, 4.6, 4.8, 4.9	5.3 / 5.4
Environmental hygiene	2.1/2.2	3.1, 3.3, 3.5	4.7 4.1-4.4	5.3 / 5.4

0.3 Notes to facilitator:

The m odules presented are participatory, which means they are designed to involve participants in decision-making, and also to use their relevant knowledge and attitudes to influence others and achieve change. The aim is to improve the health and quality of life for all members within the community. The degree of success depends largely on whether people see a need for change. If the pressure for change comes from individuals within the community rather than just from you, there is much more chance that it will take place.

The sessions or modules are set out, giving the background (introduction), purpose of the meeting, materials and time needed, the method and how to use it, relevant discussion points and notes to the facilitator on particular points   you should bear in mind. In addition examples are given of the kind of information which may be obtained from such sessions, the posters which may be helpful, the next steps to follow, and the points which should be noted down to assist in the next meeting associated with this process or topic.

The m ethods used (see list p.xv) are participatory tools which can equally be used for different types of change, so examples are given of other applications to which the session can easily be adapted. However it should not be regarded as wrong to ask direct questions sometimes, or just to discuss freely without using a specific tool. The tools are a way of assisting those (especially women) who may feel reluctant to speak out, or who have less education and ability to write, draw, or interpret pictures, but need not be used to the exclusion of all other ways. As far as possible, however try to help others to work things out for themselves, rather than instruct them and dictate what they should be doing.

However this is **not** to say that what the community says cannot be explored and investigated. If possible, participants' responses should be probed as it is important that there is clarity on **why** people feel certain practices are positive or negative. These beliefs are likely to affect subsequent decisions by the community on what they see as priority areas for change and may also offer ideas for how hygiene and sanitation can be promoted in the community.

 In particular you may find that many of the changes that you wish to bring about for health reasons, are not normally adopted for the same reasons by those who already do them.

REASONS FOR CHANGE

Ask people why they have a latrine. Very few will answer: "Because I want to reduce diarrhoeal disease". Most will say things like -:

- Because I like the privacy
- Because I don't want to be shamed with my in-laws.
- Because I don't like flies and smell near the house
- Because it shows that I am important in the village, and want to improve it

---

Ask people why they dug their own well. Few will say "so that I can wash my children more often" or "to improve the hygiene status of the house".

Most will give reasons such as -:
- For convenience (shorter distance)
- So that I can make bricks (or beer) and so make money more easily
- So that I can water plants on my own land, as well as supply the house
- So that my wife does not stray far to collect water and I can see where she is
- So I can improve it knowing that it is my own and my money will benefit me directly

---

Ask people why they use soap. Very few will say: "Because it kills bacteria, and stops the spread of faecal coliform". Most will say things like -:

- Because it smells nice

- Because it keeps my skin from drying

- Because it makes me feel clean and refreshed

Fig.02   Clean Hands – social benefits of handwashing

I feel proud when my children are admired. Washing their hands with soap makes them smell nice and clean and adorable.

If these are the reasons why people already adopt the practice you are aiming for, do not try and sell the idea **only** on the basis of blocking disease transmission, especially as many people do not accept this idea totally anyway.

Whilst **drawings/ posters** are provided (these should be put in clear plastic pockets for use in the field), in many cases members of the community will be able to draw much more relevant pictures. This process tends to provoke more discussion and can provide a record of the situation before changes are made, which can be very useful when evaluating the changes and their impact afterwards. It will therefore be helpful if you build up your own 'library' of pictures, including ones relevant to individual communities.

In general **target groups** are all members of the community, young, old, male and female, unless otherwise mentioned in the text. The idea is to get the involvement of as broad a cross-section of the community as possible, and to give all an equal opportunity to express their opinions and beliefs. Large groups (more than about 15-20) will often need to be divided up (randomly or by gender/age) to make discussion easier, and then each group presents its conclusions. Please note that in some cases, especially where water sources only serve a small number of houses ('family wells'), the community may be the whole village, if discussions are focussing on selection of sources to improve, or hygiene issues. However when planning activities, it will require smaller groups, such as those using one source.

Promoting discussion.   Ideas, opinions and plans should, as far as possible come from the participants. You, as a facilitator, are there to promote discussion and ensure that everyone gets a chance to give their point of view. You are in a good situation to be sure that all groups, (male/female, educated/uneducated, rich/poor, young/old) get an opportunity to voice their particular concerns and ideas. It is the role of the facilitator:

- to keep discussions focused on agreed topics,

- to make it an entertaining but thought-provoking time for all concerned and

- to ensure that at the end of the session all are clear what the main conclusions have been, and what is going to be done next.

A facilitator should not lead discussion and put forward their own ideas. Wherever possible, when questions arise from participants, pass the question back to the whole group to see if others can answer, before putting your own point of view. The temptation is to take over, as you may feel that your knowledge is greater, but try and support the good ideas of others and build on them rather than putting over your own ideas as the starting point. Experience shows this to be a more effective way to get people to change their attitudes and behaviour. The first few times it may be hard to act less as a teacher and more as an interviewer or actor, but when you see the results you will begin to find it easier and more rewarding!

Summarising discussions. At the end of each session ask participants to pick out what they have learnt, and what they want to do next. Summarise the conclusions yourself, emphasising the points which are going to be followed up later. Record the conclusions to each session in your note book so that you can refer to them when planning the next meeting. If possible keep a separate section for each village you are working with.

The **materials and methods** given are examples, rather than a strict procedure. They are given more to show the wide scope of the participatory approach than to dictate a specific routine which cannot be varied. In this way you will find that each time you start the process, you and participants will learn something new and gain a refreshing perspective on problems, rather than just giving out the same lesson time and again.

If people come to you with a specific problem, then it may not be necessary to undertake all the Stage 1 steps of 'Identifying and prioritising issues'. You should pick a series of tools using Tables 0.1 and 0.2 that will assist the community to solve that problem and then if they feel successful they may feel capable of tackling another one. Then you have really succeeded!

SECTION ONE

GETTING TO KNOW THE COMMUNITY AND IDENTIFYING KEY ISSUES

1.1 GETTING TO KNOW THE PLACE (EXPLORATORY WALK)

1.2 IDENTIFICATION OF KEY ISSUES (COMMUNITY MAPPING)

1.3 THE COST OF DISEASE

1.4 HOW DISEASE IS SPREAD

1. GETTING TO KNOW THE COMMUNITY AND IDENTIFYING ISSUES CAUSING MOST CONCERN

INTRODUCTION

This section helps to familiarise the facilitator with the community and allows community members to begin to focus on important issues and problems which they are facing. It is a period during which you, as facilitator, establish a good working relationship with the community. During this time you can see the way in which the community is organised, what different groups exist and to what degree all members are easily able to speak out and be listened to. This will also help you to establish whether the whole community views the same problem as a priority or whether it is only the view of one or two individuals. 'Communities' seldom act as one unit, and the degree to which various groups see problems differently will have a big effect on how successfully any problem is overcome.

The examples given allow for identification of problems which are affecting the community in all sectors, or within water, sanitation and hygiene, and then for them to decide which problems are the ones to which they wish to give priority. In that this manual relates mainly to water, sanitation and hygiene, you should make it clear that these are the topics on which you are concentrating. If the community regards other things as more important at this stage, find out from your district office which organisation can best assist with the problems identified and

> a) inform the relevant district officer of what you have found and
>
> b) inform the community of the office they should contact.

Before undertaking any of the exercises in this manual it will be necessary to introduce yourself to the community and its leaders, if you are not previously known there. In any case it will be necessary to explain the reason for your visits, which will be primarily to learn from them, and to make use of their knowledge to help them improve their situation, but may also be:

a) in response to a request for assistance from the community

b) as a result of concerns (your own or others) over unusually high incidence of certain diseases (especially diarrhoeal)

c) in response to district availability of funds for development or preventive health activities.

Always follow local good manners in obtaining permission from the head-man/ woman or chief, explaining the purpose and discussing what is the best level of involvement for traditional leaders.

1.1   GETTING TO KNOW THE PLACE (EXPLORATORY WALK)

INTRODUCTION Undertaking an **exploratory or transect walk** can be done before a community mapping exercise, to help the facilitator see some of the issues which may arise. However it is not essential, especially where you are already very familiar with the community concerned.

PURPOSE To see at first hand, with a few members of the community, the features which may affect water, sanitation and hygiene. These may include aspects which are not, initially raised in larger groups, or which may be difficult to represent on a map.

TIME 1 hour.

MATERIALS Note pad and pen

METHOD

1. Walk around the village with a selected group of three or four people. Visit any water sources, and houses which are regarded as the best kept and less well kept. Observe as you walk both the behaviour of people and the environmental health status of the village, but do not comment on it negatively.

2. Ask those walking with you how they view these aspects of the community, and what problems they identify as they walk around.

3. Take the opportunity to talk to people who may not attend community meetings such as the elderly and youths.

4. Make notes on what you observe and what you learn during the walk, especially the views and opinions of others.

NOTE TO THE FACILITATOR:

- Do you get the impression of a community which takes pride in its surroundings and way of life, or does this need to be developed?

- Are there human faeces found on the ground, and is there rubbish indiscriminately thrown around, or does it appear that most people have standards of behaviour which include safe disposal of faeces and other solid waste?

- Are water sources well kept and are people's practices in collecting water likely to keep water clean or lead to contamination?

- Do most people have latrines, bathhouses, pot racks and rubbish pits, or are these rare / non-existent?

- Are there signs that the community is relatively well off, or particularly poor? (Does house construction suggest that some people have sufficient resources to buy cement and employ builders? What indicators are there of sources of income?)

- Summarise in your notes, the issues which you see as being most likely to affect health/well-being, and those remarked on by others. This can be compared with what comes out of the community mapping which involves a larger number of people.

RECORD KEEPING

Keep a record of your first impressions of the community, and of the issues brought up by those you talked to. Explore some of these further during community mapping (see 1.2)

Example: Use of Exploratory Walk

The Environmental Health Technician from Mwamba Rural Health Clinic (Mbala) visited Kawimbe. He walked around and talked to people. He observed that houses were brick built, well kept and had latrines, suggesting that people took pride and had some resources. However, the unlined well close to the houses was dirty and had no protection. Chickens sat around it and dust and run-off could enter the well. He was told that drinking water was therefore not taken from this nearby well, but from a distant spring, which was tiring for the women. His interest in the situation prompted the headman to begin to think of improving the source and they began to work together to see if this was something that other users wanted too, and that might receive some assistance from outside.

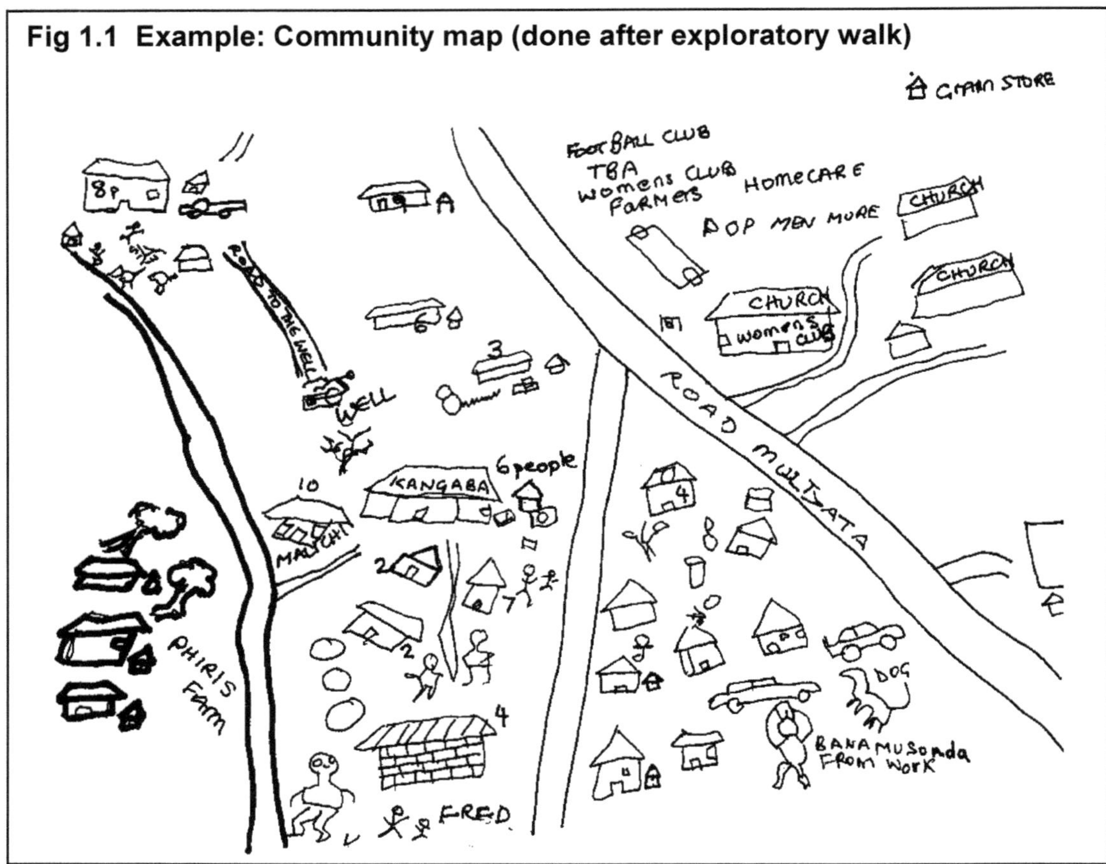

Fig 1.1 Example: Community map (done after exploratory walk)

Exploratory walk notes: Multyata

1. Several big houses, some with zinc roofs and brick walls. Two households with a car so it is quite a rich village.
2. The path to the well is clean and grass is cut, the surroundings are clean
3. There is a football pitch and several churches and associated   clubs, suggesting the community is strong and active.
4. Most houses have grain stores and people are hard-working.
5. About half of the houses have no latrines, and dogs are found where people are defecating. Pots are kept in the houses, not on the ground.
6. The latrines are mostly quite bad, with roofs falling in and grass walls rotting. People say the pits collapse too easily
7. People at the well say they want to make it clean and stop rain falling in. At present they go far for drinking water and only use the well for washing, and bathing. Each person brings a bucket, and the well belongs to the headman. It has not dried since it was dug. Ropes lie on the ground while people rinse out their buckets, and water is ponded round the well.
8. Children are mostly clean, and well –dressed but their hands are dirty.
9. Several houses have kraals for cattle and boys/men from other (poorer?) houses take the cattle to pasture each day.

1.2 IDENTIFICATION OF KEY ISSUES (COMMUNITY MAPPING)

INTRODUCTION At this stage it is important to establish a good working relationship with the community, and a situation in which all members have a chance to contribute their ideas and opinions. The exercise of **community mapping** allows the discussion of a broad range of issues, and the identification of positive aspects of the community as well as the problems. It is a very important exercise, which can be used in many different ways, and at various stages in community development.

PURPOSE

To enable participants to identify important issues and problems in their community. These may relate to broad development issues or be more specific to water, sanitation and health. It is also an exercise to help build team spirit and pride in what they are able to achieve.

TIME 3 hours, but other aspects may be added later

MATERIALS

Pens and flip chart paper. A4 paper for facilitator to make a sketched copy of the map, and notes on issues raised. If there are no large sheets of paper, the map can be drawn on the ground using local materials to represent features such as houses etc.

METHOD

1. Greet the community and introduce yourself and the purpose of the exercise.

2. Community mapping is being carried out more and more often for a variety of different purposes and therefore it is necessary to check -:

   a) if a map already exists
   b) if so, for what purpose was it drawn and what conclusions were reached
   c) if people would prefer to make a new map, or add to the one already made.

3. If numbers of people are sufficient, ask men and women to draw different maps. It may be that a group will ask a school child to help them, especially if one is known to be good at drawing. This will be helpful as the group will then talk about what they want drawn and why.

4. Motivate the identified group from the community to participate in map drawing by telling them that they are experts as they know their village better.

5. Ask the group to choose where they would like the map to be drawn.

6. Help the group to choose any symbols for different features e.g. houses, toilets, wells, utensil racks etc, but make sure that a key is made to explain what each symbol means.

7. The map should initially include the following :

   - houses with their latrines, bathhouses etc.
   - water sources/ streams & rivers
   - roads, any special features such as health centres, schools, churches
   - any other features that participants feel are significant features of the village.

8. Summarise what has been learnt during the mapping, with the participants, and ask them to bring out those problems which the mapping helped to identify, and which they may wish to try and solve.

DISCUSSION POINTS

- Are there any problems which people feel are affecting their health and well-being?

- What aspects of the community would they most like to change in order for their quality of life to improve. These can be listed rather than prioritised at this stage.

- Into what different groupings could the community be divided? Do different people have different views as to what the main problems are?

- What are the aspects of the community about which participants are proud? Can these be built upon to bring about positive changes in the community?

- What aspects cause them embarrassment?

NOTE TO FACILITATOR:

1. Try not to give your inputs to the map, but let it come from the groups themselves. You may be surprised at how good people are at mapping and drawing!

2. It may take some time for the map to be drawn, but it is a good investment of time as the process and the finished product are both valuable, and will be referred to many times in the future. Record relevant aspects of the discussion as you watch them drawing.

3. The mapping exercise can be used in conjunction with other participatory tools such as focus, group discussion, semi-structured interview, ranking and history line.

Results of community surveys in Zambia

When community mapping and seasonality calendars were first used by Environmental Health Technicians in a baseline survey, they all remarked upon the change in their relationship with the community, even when they had already been working in the same area for several years. They became more aware of the concerns of the communities, which were not necessarily the ones that they themselves had thought were most pressing. They also felt part of the team solving the problem, rather than an outsider on whom everyone was depending.

Example of use:
Problem identification in sanitation

While drawing a community map, some participants at Sikuyu village near Mongu, Zambia, complained that whilst they had made the effort to build a latrine each for their families, their neighbours had not. The flies and smell of faeces around them troubled them and they though the neighbours were lazy. It became clear from the map that those without latrines were mainly households with orphans or with only elderly people. At the same time, some of those with latrines were found to have put them too near the water source. The community requested that some owners relocate their latrines and for people to help provide materials to those unable to afford their own. They identified on the map which was the best-built latrine and asked the owner to advise everyone on how he had done it, and what measures the family took to keep it nicely. That owner was proud to be a model for others to follow, whilst others were embarrassed to be regarded as lazy, and quietly went away to improve the situation.

Fig.1.2 Map of Kandiana village

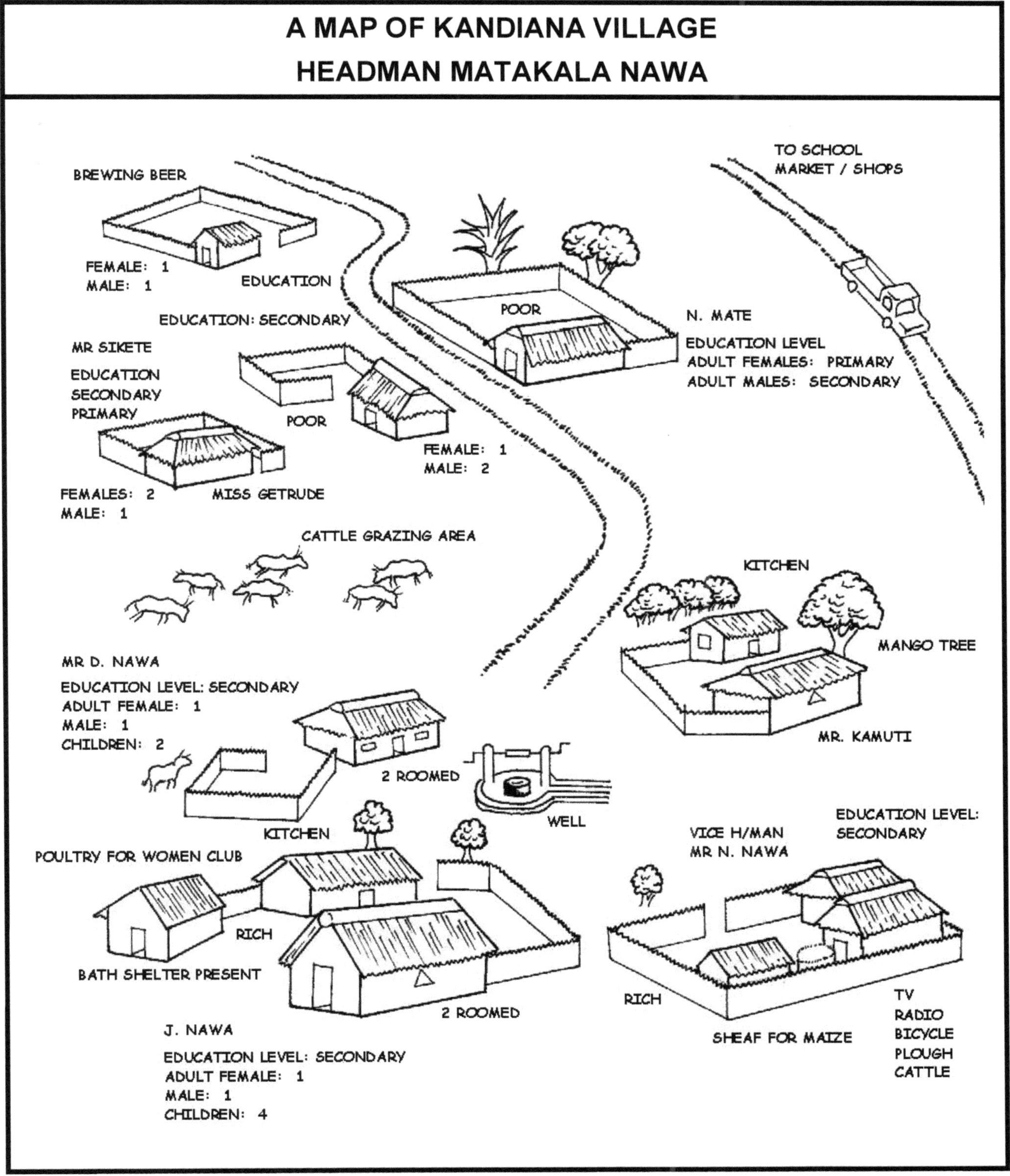

Other applications

Community mapping can also be used for:

- Planning (community draws a map visualising how they would like it to be)

- Evaluating (community prepares the initial map with one showing changes)

The map can also be used for aspects such as wealth ranking, various social groupings, identification of skills and fund-raising potential (See Sections 3.4 and 3.5) establishing model households and community self-monitoring systems.

Fig. 1 .3 C ommunity mapping can be an a   bsorbing act ivity which throws up lots of ideas, and gives everyone a chance to be involved.

1.3 THE COST OF DISEASE

INTRODUCTION

It is often accepted as an every day situation for children or adults to have sickness such as diarrhoea or malaria. What differences would it make if people were able to prevent it? This exercise looks at the inconveniences and costs of being sick with diarrhoea. The same method can be used to highlight the problems of other diseases in order to create awareness and assess demand for change. It uses role play and focus group discussion.

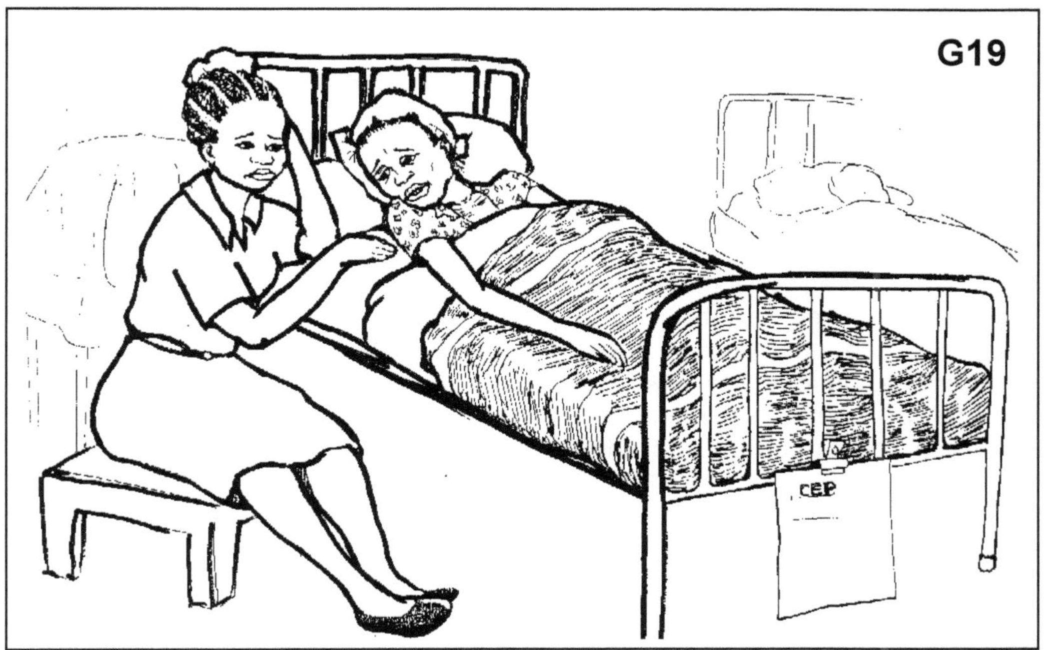

Fig 1.4 The costs of sickness are not just that you are out of action or have to buy drugs.

PURPOSE

This session can be used to help people see how much of a problem diarrhoea can bring, especially to children. It therefore emphasises the benefits that can come from any efforts to reduce the incidence of diarrhoea, encouraging people to take more care. It is followed by sessions which identify in what ways the risks can be reduced, if there is a strongly expressed demand to improve the situation.

MATERIALS

Pen and paper to record main points and conclusion.

If required, posters of daily activities may be used to structure discussion.

TIME 1-1½ hours

METHOD

1. Before the meeting ask 5 or 6 people if they would be willing to act in front of others. Explain that you want to highlight the inconveniences and longer-term problems which sickness in children may cause. Ask them to prepare a small play, and offer any help they may want.

2. Divide other participants into 4 groups, men and women, adults and children. Ask them to watch the role play and then each group to identify the key 'costs' for each group from the play (cost = inconvenience, extra work, extra things needed, problems at school or work, as well as any cash or other payments). Have they

experienced any other problems not shown in the play, as a result of this sickness in their households?

3. Summarise the conclusions and obtain participants' views on whether diarrhoea causes sufficient problems in the community to justify trying to reduce it.

NEXT STEP

If diarrhoea is identified as a real problem needing solutions, go on to Module 1.4

BACKGROUND

In 2009 27% of children were moderately or severely malnourished in sub-Saharan Africa. Malnourished children are more prone to diarrhoea, do less well at school and tend to be weaker. These are problems which are carried on into adult life, if the child survives. However 1 in 8 children in the region died before the age of 5, and diarrhoeal disease is the second leading cause of death. Diarrhoea both results from and leads to malnourishment. Thus water, sanitation and hygiene are major elements in child survival, and safe water availability can play an important part in improved nutrition, through its productive use and its role in hygiene. (statistics from UNICEF Childinfo 2009)

DISCUSSION POINTS

- What are the direct effects of a child being sick?

- What are the effects on others sharing the same household?

- If children are frequently sick with diarrhoea are there effects which carry on into their adult life?

- If the mother or father is sick, what effect does it have on others in the household?

- If the person did not have to be sick, what benefits would there be?

- Do people think it would be worth trying to improve the situation if it is possible or is the level of diarrhoeal disease not felt to be a problem in the community?

Demographic Health Survey Zambia 2007

In Zambia the DHS survey for 2007 shows more than half of children under 5 (56%) had had diarrhoea and oral re-hydration (ORS) in the previous two weeks. In contrast, for Ethiopia the comparable figure (DHS 2005) is 15%.

RECORD KEEPING

The facilitator should record those aspects which participants identified as being the main 'costs' of diarrhoeal disease.

Results might be recorded under headings such as:

- **Economic** (e.g. cost of Oral Rehydration Solution / traditional medicine, loss of work time)

- **Social** (e.g. smell, infecting others, no time to stop and talk)

- **Quality of life** (e.g. tiredness, having to collect more water, clothes getting worn away from washing, other family members disturbed)

- **Health** (e.g. more open to other illness, drugs less effective when purging, lose weight and strength)

- **Other** (e.g. can't concentrate at school, other children laugh when one has to leave class hurriedly, lack of local herbalist means long journey for treatment)

EXAMPLE OF ROLE PLAY

At Liyoo village five players volunteered. They acted different incidents from the middle of the night to the middle of the day, first part at home, second at school, as follows:

1. Young school-age child 1. Wakes in night with diarrhoea, wakes others with crying, frightened of the dark and path to the latrine, can't sleep, keeps others awake. Doesn't go to school.

2. Mother. Tired from lack of sleep, extra clothes to wash, more water to collect, goes to find local herbalist who is out, plans to go to clinic if child not better next day. Long distance means she will get behind with her other daily chores.

3. Father. Discusses with wife if child needs treatment, and how they could pay. Gets angry because of lack of sleep and finds fault with wife. Blames her for the sickness, for no warm water for him to wash, and food not ready.

4. School age child 2. Tells teacher that brother/sister is sick so not coming. Child 2 is sleepy in class.

5. School teacher. Complains that child is often off sick and is getting behind with work. She has no school books so cannot give him work at home, and she has too many pupils to have time to help one catch up. The sick child may have to do the year again, but should be finishing so can help parents in the fields, or go on to secondary school.

Fig 1.5 Role play can provide entertainment and start people thinking and discussing.

The cost of treatment.

In 1998 the average cost of a visit to the doctor in Zambia was over 16,000 kwacha (US $7), and for a traditional healer 10,000 kwacha (US $4). A visit to the clinical officer was cheaper at around one US dollar. This was in areas where over 64% of the population lived on less than US $1.25 per day.

1.4 HOW DISEASE IS SPREAD

INTRODUCTION

Consideration of water, sanitation and hygiene relates mainly to reduction in diarrhoeal diseases. This exercise involves discussion and sorting out of pictures that show how germs causing diarrhoea are passed from one person to another and actions that can be undertaken to prevent it. In Sections 2.2 to 2.4 these aspects are looked at in more detail.

PURPOSE

This tool is used to assess individuals and households understanding of risk behaviours associated with transmission of diarrhoeal diseases

It helps to identify possible measures that may be put in place to prevent the occurrence of diarrhoeal diseases, and those which the community would like to tackle first.

MATERIALS

Posters (Fig 1.6) of "F" diagram (Fig 1.7), paper, pens, markers

TIME   1- 2 hours

TARGET GROUP

Everyone in the community. If possible have separate groups for men, women, youths and children, as this will help to identify different beliefs and attitudes by gender and age.

METHOD

1. Break participants into small groups.

2. Give the participants a set of pictures (F diagram) to include fingers, food, faeces, fields, fluids, flies and person as a host.

3. Ask the group to discuss, agree and show using arrows and pictures, how germs that cause diarrhoea can reach the host. Identify barriers to the faecal-oral route using cards to block the appropriate routes.

4. Capture this on a big piece of paper and present to the whole group.

5. Let the participants prioritize the interventions by ranking them according to which barriers they would like to concentrate on first, and use this as a basis for more detailed analysis of the problems. Ranking can be just putting the pictures in the order in which they wish to consider the barriers, or it can be done by scoring using '**pair-wise ranking**' (taking two pictures and deciding which is more important, and then the more important and one other and doing the same again, until all are in order). Allow equal ranking where it is suggested.

Some posters which could be used, but get participants to draw their own if possible.

H1, H4, H8-14, F posters (Fig 1.6)

Some of the sanitation posters including **S11-15, S2** and/ or **S8**

Good and bad water source (for example **W4, W6, W8, W12** and **W13**)

Other applications

The same system can be used for looking at barriers to intestinal worms, malaria and schistosomiasis.

DISCUSSION POINTS

- Are diarrhoeal diseases a problem in the community?

- Who are the ones who suffer most from it and why might this be?

- What causes diarrhoea?

- Why is it dangerous?

- Discuss the different ideas people have and then ask them to use the pictures and explain how germs may move from one host to another.

- What actions could block this transmission?

NOTE TO FACILITATOR:

- Identify people's perceptions of risk behaviours, and their reasons.

- Establish differences in levels of understanding by sex and age.

- List or ask people to make drawings of the main risks identified.

- Let participants prioritise activities by ranking them according to which they consider as the most important to tackle first

- Make a note of all gaps identified and use this for planning and selecting the steps which might follow on, depending on the priorities identified

---

Example of use

It is commonly found worldwide that children's (fresh or dry) faeces are not regarded as dangerous. During a session looking at how disease is spread, Sefula Rural Health Clinic explored why this belief was held, rather than telling people it was wrong. Participants identified that children tended to be sick more often than adults, and washed less often, but that they ate the same food and drank the same water. They began to discuss whether it was then reasonable to believe that only adult faecal matter was dangerous. They discussed how easily children could come into contact with faecal matter and how mothers could increase or reduce this risk. More families began to bury children's faeces or put them in a latrine.

RECORD KEEPING

The facilitator should record the main problems identified and the barriers which the community feel should be focused upon first.

Table 1.1 Barriers to spread of disease, and steps in establishing them (see also Table 0.2)

NEXT STEPS. Choose modules to fit priority problems/ barriers identified.				
PRIORITY CONCERNS	**PROBLEM ANALYSIS**	**COMMUNITY PREPARATION**	**DEFINING OPTIONS**	**ACTION PLANNING**
Faecal disposal	2.2	3.1-3.5	4.3,	5.3+5.2
Safe water supply	2.2	3.1-3.5	4.1, 4.2, 4.4	5.1+5.2
Handwashing/ personal hygiene	2.3-2.5	3.2+3.3	4.5, 4.6,4.8, 4.9	5.3 / 5.4
Environmental hygiene	2.1/2.2	3.1, 3.3, 3.5	4.7/ 4.1-4.5	5.3 / 5.4

Fig 1.6 Hygiene Posters (for use in 'F' diagram)
FAECAL- ORAL TRANSMISSION and HOUSEHOLD HYGIENE

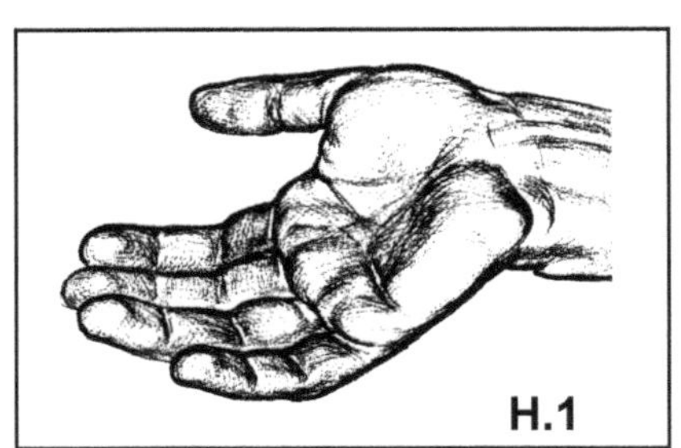

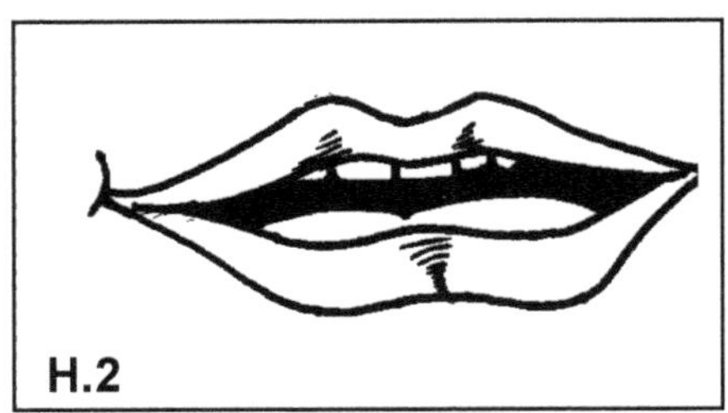

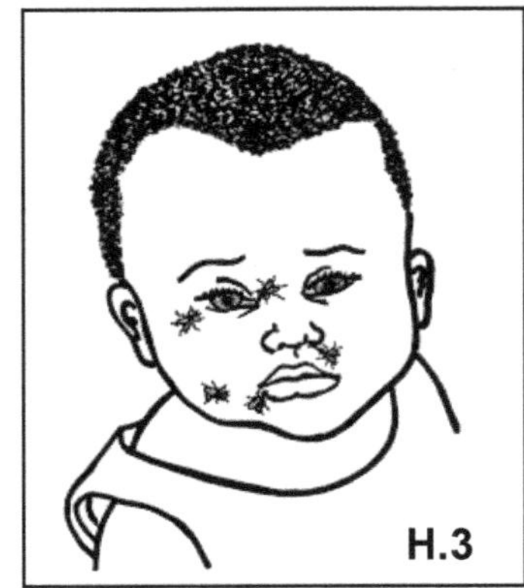

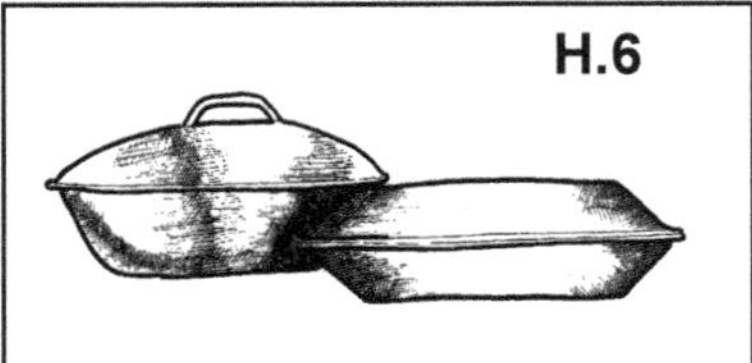

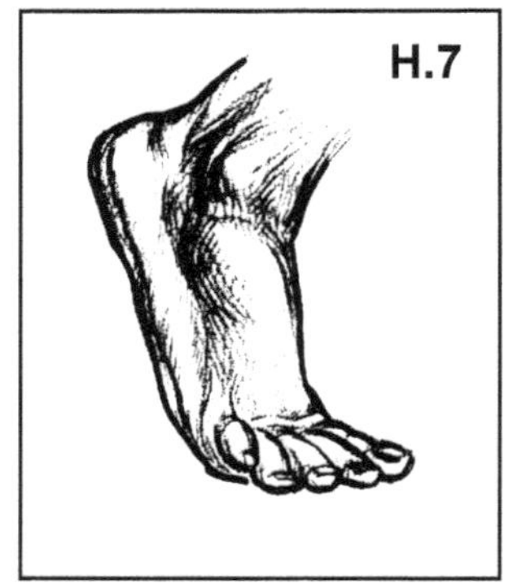

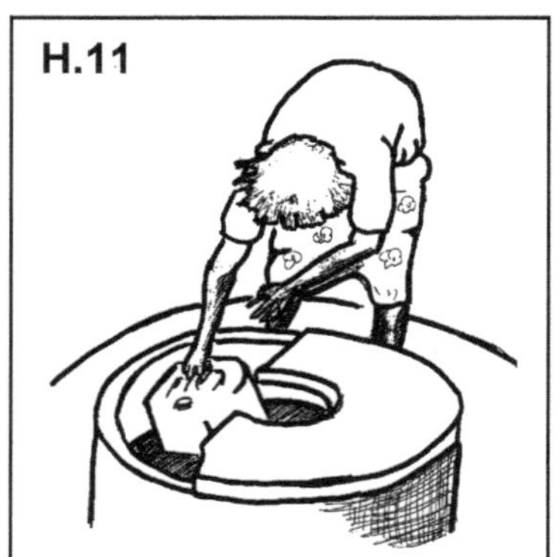

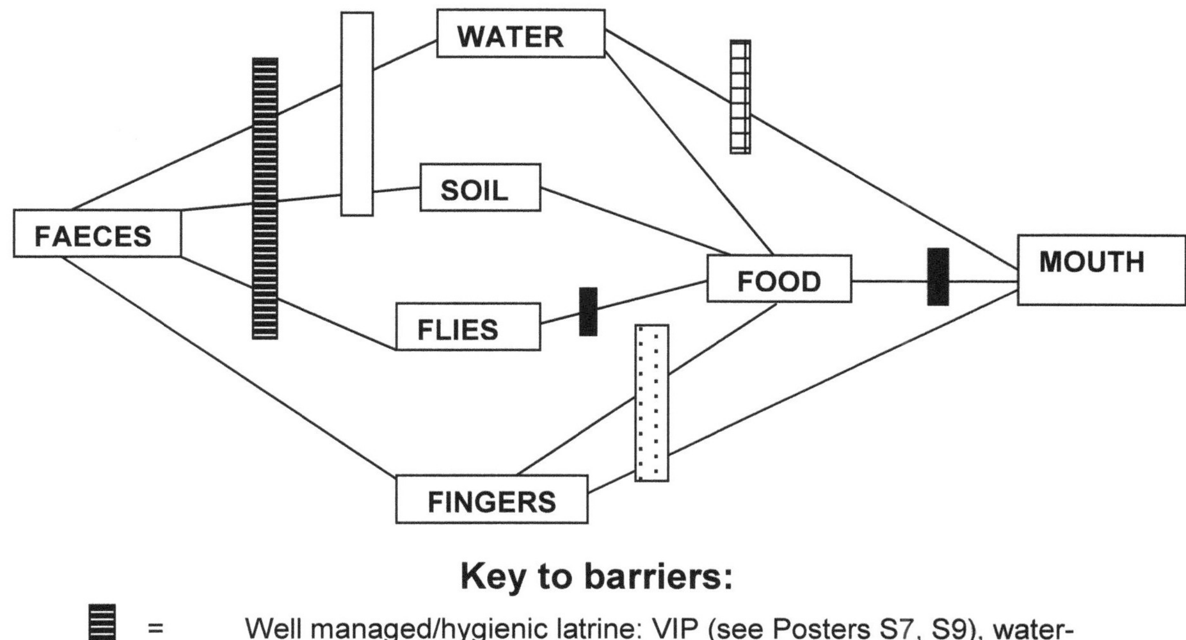

Key to barriers:

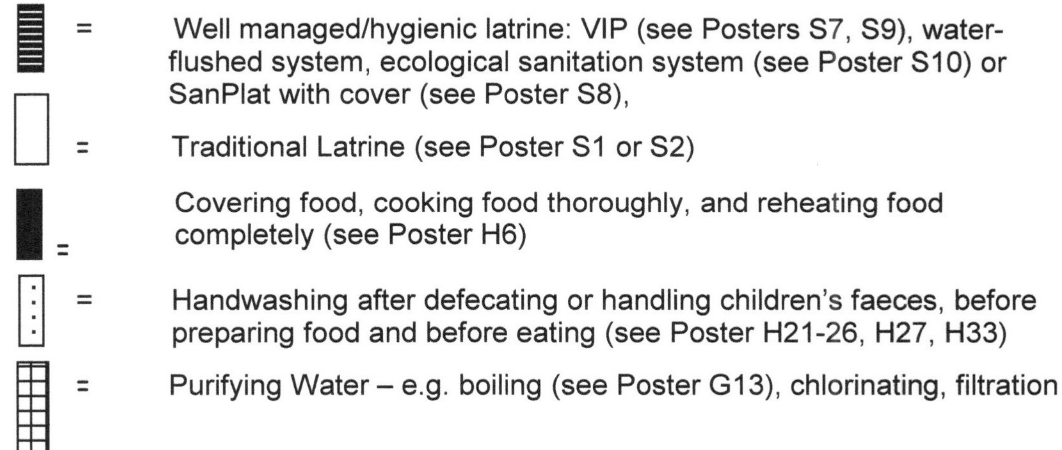

= Well managed/hygienic latrine: VIP (see Posters S7, S9), water-flushed system, ecological sanitation system (see Poster S10) or SanPlat with cover (see Poster S8),

= Traditional Latrine (see Poster S1 or S2)

= Covering food, cooking food thoroughly, and reheating food completely (see Poster H6)

= Handwashing after defecating or handling children's faeces, before preparing food and before eating (see Poster H21-26, H27, H33)

= Purifying Water – e.g. boiling (see Poster G13), chlorinating, filtration

Sanitation alone will not have the desired impact on community health in the absence of effective behavioural practices that further reduce the likelihood of disease transmission.

So, a "sanitation project" will have little impact on household health, despite a new latrine if:

- Some people (such as children) do not use the latrine

- Children's faeces are not safely disposed of

- Latrines are not properly maintained or cleaned

- Hands are not washed by all at critical times (after defecation, after cleaning a child's stools, before eating and before preparing food or drawing water)

- Water source is not protected or water purified prior to consumption

- Food is not prepared properly (cooked thoroughly, reheated thoroughly, stored in a way that does not allow for it to be contaminated by insects or small animals, or cleaned thoroughly before eaten raw)

SECTION TWO

ANALYSING PROBLEMS AND ASSESSING UNDERLYING CAUSES

2. ANALYSING PROBLEMS AND ASSESSING UNDERLYING CAUSES

INTRODUCTION

This stage involves considering both the physical and behavioural problems relating to water, hygiene and sanitation. The problems identified can then be discussed in 'Planning for solutions' (see Sections 4 and 5) to define the changes needed and what they would require.

It is well known that water-related diseases are often more closely linked to inadequate sanitation and people's hygiene practices (the primary causes of faecal contamination) than they are to unsafe water supplies. Finding out people's attitudes and practices, and the reasons they consider certain standards of behaviour acceptable or unacceptable, forms a first step in encouraging people to change their habits, by creating and/or assessing their demand to do so.

It is important that these modules are used to find out what **members o f t he com munity** think and believe is good and bad about their current situation. Great care must be taken to avoid using the session as a hygiene education meeting in which the community are told which of their practices is 'right' and 'wrong'.

Its success depends largely on whether people see a need for change. **If t he pressure for change co mes from i ndividuals within the community rather than from you, there is much more chance that it will take place.**

In many cases a community may seek assistance having already identified the problem they wish to solve. In that case this section may be omitted or used in a way to explore more of the underlying reasons that such a problem exists.

Five examples are given of how to investigate and identify root problems. Since the same tools can be used for many purposes, other applications are also suggested. We hope that you will be able to feed back other examples of how you have used the tools, and tips on what topics are raised and what methods gave the best results.

Each session is likely to produce a significant amount of important information related to people's practices, perceptions and beliefs. It is important that as much as possible of this information is recorded as it can be used in subsequent planning, promotion, monitoring and evaluation work. Just as important as the 'answers' to the exercises are the comments made by members of the community during the discussions that produce the final output for each tool. The reasons they give for adopting good or bad practice should be built into any plan to promote change – this is even more important than the theoretical reasons you know from your training.

2.1 WHAT IS WRONG WITH OUR WATER SUPPLY?
(This can refer to any type of technology.)

PURPOSE   To get users to identify those aspects of their water supply which cause them most problems.

MATERIALS

A4 paper, (flip chart paper for mapping), coloured pens and pencils, eraser.

TIME 2 hours

METHOD

There are several ways to promote discussion at this stage. (eg. Role play as in 1.3, blind voting with identified problems as in 2.2). Choose the ones which best suit the problems raised during community mapping and discussion of health issues. In this example **focus group discussion** is combined with **mapping or drawing.**

Rather than using posters of other sources which may not have relevant features, it may be better to get the community to draw its own. In this way many relevant features may be identified which would be omitted in standard drawings. If there are seasonal problems ask for a drawing representing the wet season and dry season situations. If several sources are used, get those using the same sources to work together on the drawing.

1. **Understanding the problems.** Drawing of the source. Where does the water come from? What may affect the quantity of water available or the ease of access? What may affect its quality? (use drawing and community map). − Are there any sources around which may affect it?

2. **Identifying what could be changed**. Add on the drawing (if not there already) the features and activities which are thought to affect quality and quantity. Which of these features could be reduced or removed if rules were made and kept to? Which require changes to the source, or repairs?

3. **Summarising the conclusions.** The drawings or map may need to be supplemented with a few notes on aspects which may be difficult to represent by drawings (but try!) (eg ownership / management, lack of resources)

Example of results 1

Users of a scoophole identified the problems that arose because leaves fell into the water, people stepped into the water to collect it, and even animals were drinking from it. When the water level fell in September they had to move to a more distant source that was better protected.   They liked their own scoophole because it was near, tasted good when it was clean, and saved them from the embarrassment of using other people's water supply. They felt they would like to improve their own supply rather than contribute to one they regarded as belonging to other people, and agreed to meet again to look into how this could best be done.

Example of results 2

At William Chibwe village the community identified seven sources being used. They decided that they would try and improve the two which provided the most reliable supply and were used by the most people. They felt that run-off entering the well, dirt being blown in, and the use of many different buckets for drawing water caused problems which they could solve. They respected the owners of the wells which provided water all year and wanted to find ways to assist in improvements, while acknowledging that ownership and decision-making remained mainly with the family on whose land the wells lay.

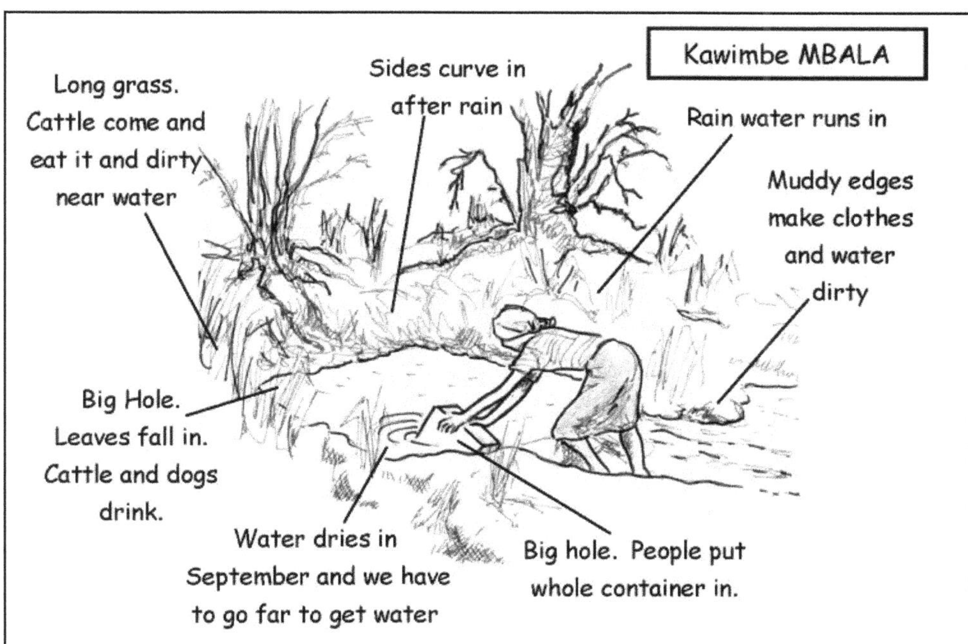

Fig 2.1 Problems identified by Kawimbe village

DISCUSSION POINTS

- **Sources of pollution identified?** (eg. latrines, cattle pens, soaking pits for skins or cassava, abandoned wells or springs, drainage canals, run-off entering the well).

- **Were any features likely to affect quantity of water av ailable?** (Examples for traditional sources - seasonal seepage points, interference with spring flow by diversion or blocking, collapsing of soft ground, growth of 'thirsty' vegetation such as eucalyptus upstream of spring or near well.)

- **Were there any practices in water collection, or water us e near the source which were felt likely to cause pr oblems to t he suppl y?** (eg. people stepping in water, putting jerrycans or buckets directly in the water, children washing in the source or playing around it, too many users (queuing), brickmaking causing dirty water to be ponded near the source, animals watered or wallowing in the drainage water etc.)

- **Are t here an y under lying reasons for t hese pr oblems?** (poor management, lack of user feeling of responsibility / ownership, lack of funds/ resources)

- **What ar e t he posi tive aspects of the suppl y?** (eg. taste, ease of access, low cost, feeling of ownership, ease of maintenance, well-established management)

NOTE TO FACILITATOR:

1. Ensure that positive features are also identified so that these can be retained / built upon as far as possible.

2. Retain drawings (preferably within the community) and go over the conclusions of this session before the next one, which will identify available resources or options for change.

Other applications

- Drawing of the 'before' situation can later be used to evaluate changes

- Drawing of a protected source and its component parts can be used to identify repair needs and management problems

- Drawings by users of what people regard as the features of a 'good latrine' and those of a bad one, can be used to help with the action plans above

- Mapping of areas of accumulated rubbish, ponded water, mosquito or other infestation can be used to plan environmental sanitation improvements

---

The next steps

- Looking at what resources the community can call upon (skills, materials, funds)

- Discussing the options of what changes can be made, and deciding which ones to undertake.

NEXT STEPS

If the community want to make changes but feel they cannot cope / do not have the resources	Go to Section 3 and select relevant modules
If the community feel capable of making changes	Go to Modules 4.1 and/ or 4.2

Fig 2.2 Source needing improvement: the same process can be used to identify why a protected source like this is in need of improvement.

2.2 DO PROBLEMS ARISE FROM HOW WE GO TO THE TOILET?

INTRODUCTION

In blocking transmission routes we saw that contact with faecal matter was the cause of diarrhoea. Communities are often aware of difficulties caused by poor sanitation, but the difficulties they identify may relate less to ill health and disease, and more to other factors such as offensive smells, lack of privacy, status or convenience, embarrassment etc. Improved ways of disposing of faeces, for whatever reason, is likely to have a positive effect on health and quality of life This module identifies sanitation practices, and attitudes towards them, so that the need for change and ways of achieving changes can then be discussed. If time is short, you may wish to omit discussion point 2 and just work on the 'where and why' aspects of defecation practice.

PURPOSE This session aims to clarify what people currently do about going to the toilet. It also looks at what they feel is good and bad about what they do now, what problems are caused and why they think in this way. They will then be in a position to consider what changes, if any, they would like to make.

TIME 2-3 hours

MATERIALS

Posters showing different defecation practices; posters depicting different times of the day; plastic folders; counters or stones; paper and marker or drawing pens.

> Suggested posters, if required, could include -:
>
> Different sites for defecation (S1-17)
>
> Different times of day/ activity (G1-14)

TARGET GROUP(S)

Community group as a whole; Separate groups of men and women (and children if sufficient numbers).

METHOD

1. Explain the purpose of the meeting. and arrange groups of men, women and, if sufficient numbers, children also.

2. Ask people to discuss at what times of the day (or night) different members of the family go to the toilet (to defecate). After a short discussion time show them a set of activity sheets and ask each group in turn to select activity sheets which are most relevant to their group, and add any more that are mentioned. They may include:

   - First thing in the morning
   - Last thing before going to bed
   - Middle of the night
   - Before going to collect water
   - Whilst in the fields / herding animals
   - After / before eating
   - Whenever they need to go!

3. Ask each group to discuss the different places where people (including small children) go to the toilet, without individual respondents indicating whether they use this method or not. Select sanitation ladder pictures to represent their answers, and ask them to draw any additional options not already represented by drawings.

4. Set out the posters on the ground, away from where people are sitting so that people cannot be seen when they are voting. Form a matrix with times of the day along the top and the types of practices down the side (see Fig 2.3 p. 31, for an example of a matrix). Explain to the group that the purpose of the exercise is to see, in general, what people in this particular

community do at present, and that they should vote according to what they did the previous day. (If time is short, leave out the time of day, and vote simply on what type of practice they did yesterday.)

5. Demonstrate two examples so that people see what to do, and offer to assist those unsure how to place votes. Individual members are then asked to place counters or stones in those 'pockets' in the matrix that correspond to their own practices.

6. Each voter may need more than one counter/stone depending on how many times they went to the toilet during that day. If there are sufficient participants (more than 15) men and women can be given different coloured or shaped counters/stones so that it can be see if the practices of men and women differ.

7. The votes can be counted publicly and the results discussed without the embarrassment of identifying the practices of individuals.

8. Discussion can be assisted by asking the groups of women and men to sort posters of 'where people defecate' into what they regard as good, bad and 'in between' practices and to explain their reasons. Ask the women to present their conclusions and then ask the men to comment on any differences in their opinion.

9. Summarise the conclusions and ask participants to vote whether they feel there is any need for change. If more than 50% say 'Yes' offer to return to discuss in what ways they would like to improve the situation and how this could be done.

DISCUSSION POINTS

- Which times/practices are most common; why ?
- Do men, and women have different practices/times; why?
- How do the practices of children (young/old, boy/girl) differ from adults?
- Which practices cause problems to the person concerned or to others and what are the problems?
- What things stop people changing to different practices?

NOTE TO FACILITATOR:

Many people are reluctant to discuss what they do about going to the toilet and are embarrassed to talk about such things in public. Therefore this session should be run sensitively and with respect for people's feelings on this issue. Some of the methods suggested should help to allow people to 'say' what they really feel since they are, in part, 'secret', coming from the group rather than the individual.

Avoid telling the community what is right or wrong about their practices, but encourage them to come up with these ideas for themselves.

NEXT STEPS	
Discussing the options, agreeing changes/ technology and behaviour	**Modules 4.3-4.9**
Looking at the resources the community can call upon to make these changes	**Modules 3.1-3.5**

RECORD KEEPING
You should make a note of the matrix made, including the results of how many men, women and children vote in each square (category). Also note down special reasons people give for **why** they do what they say they do. Also note if there are groups who are not feeling free to participate in the discussions and see if it is possible to talk to them separately.

Example from Choobana village in Monze

During focus group discussions with the residents of Choobana village in Monze district the Environmental Health Technician found out that only around half of the households had a latrine, which meant that the remaining 50% of households used the bush for their toilet. The people from the village believed that this allowed flies to sit on the faeces and then to travel and sit on the food that people ate. They also said that people did not wash their hands after going to the toilet which could allow 'dirt' to be transferred when people shook hands when they greeted each other. These practices were seen as problems because the community associated them with people getting diarrhoea and dysentery, which were identified as two of the most common diseases in the village, being particularly prevalent during the dry season when water was scarce. The community felt that there might be some immediate things that they could do to improve this situation and agreed to meet with the Environmental Health Technician again to discuss what these changes would be.

Table 2.1 Example of matrix: times of defecation

When? / How?	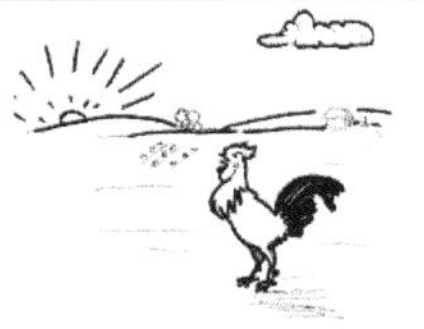	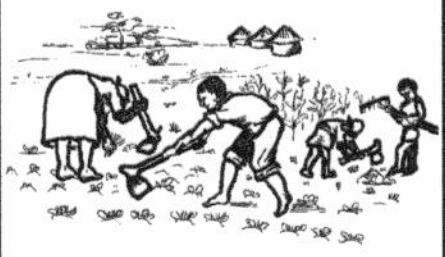	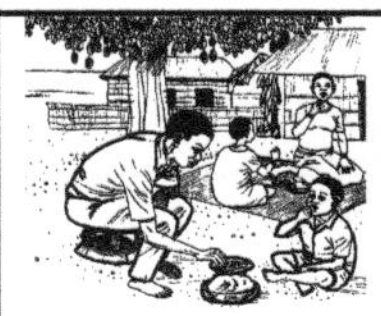	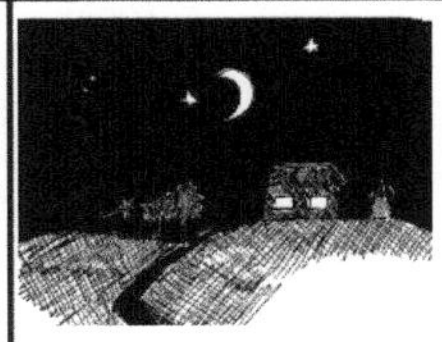
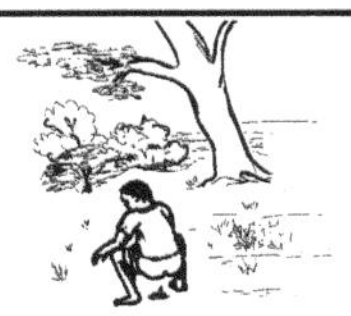				
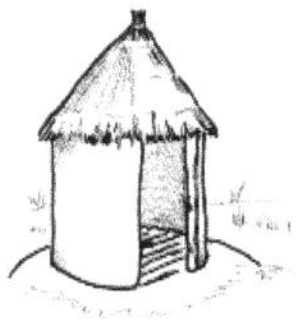				
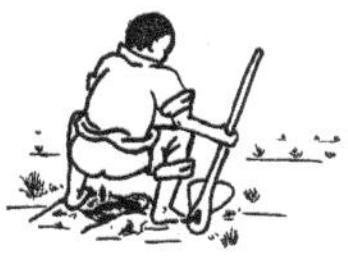				

TABLE 2.2 EXAMPLE OF PROBLEMS BROUGHT UP BY TRADITIONAL LATRINE OWNERS

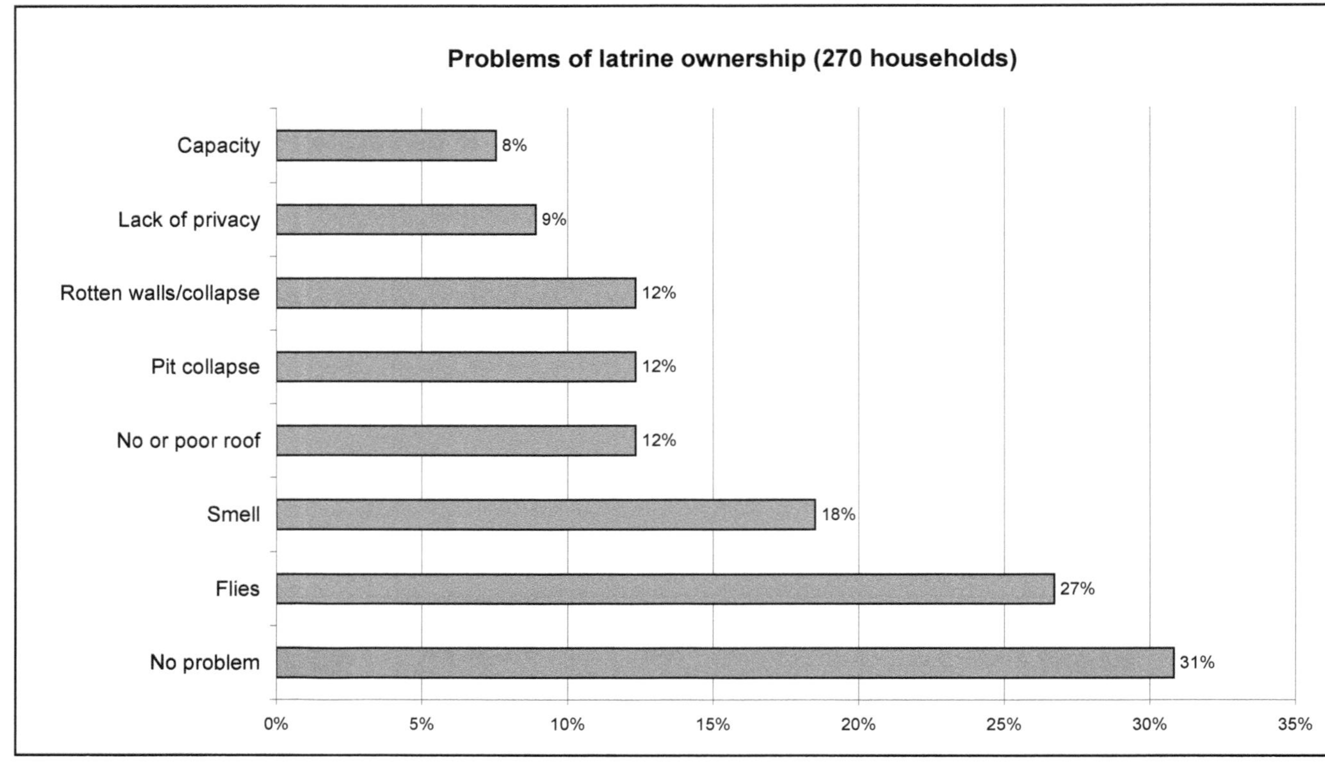

2.3 CAN HANDS REALLY TRANSMIT DISEASE?

INTRODUCTION

In blocking transmission routes, fingers, fluid and food were some of the ways disease was found to be spread. This module explores further the ways that disease is spread by dirty hands, through **role play** and **focus group discussion**. It combines activities which may be identified from the daily activity analysis, with a demonstration of how easy it is for dirt to be passed from one person to the next, and how even hands which look clean can have harmful dirt on them.

It also emphasises how much more effective washing is when done with soap.

PURPOSE To show the faecal-oral route of transmission, and to highlight the importance of washing hands with soap after touching faeces

MATERIALS

Oil, crushed/powdered charcoal, water container, cups, dishes, large bowl for hand washing, pouring cup, soap. Try to obtain items that will show up the charcoal. Garlic, onions or chillis may also be used to show how hands may look clean but still be dirty/smelly.

TIME 1-2 hours

METHOD

Sketch / play with interaction from the audience. Members from the community/ committee perform an interactive play using props.

- Begin with a sketch in which a community member goes to the latrine where he/she puts an amount of oil on their hands, then covers them with powdered charcoal. Explain that this represents the dirt which tends to get on hands, however careful one is in cleaning oneself.
- The actor then returns greeting people he/she meets (who in turn greet others), drinks from the cup (use a white one or glass), has dinner with family washing in the communal bowl, eldest male to youngest female. Let people see how the dirt accumulates in the water, which is used by the next person.

- Do the same exercise of putting charcoal and oil on the hands of a woman, who picks an empty water container, and takes it to the source (the bowl filled), and puts the container in the source, and/or gives her husband a cup of beer, or other activities people suggest.

- Compare the amount of charcoal/oil left on the hand after washing briefly in water, and after using the same amount of water with soap.

- Through this process the audience become familiar with faecal-oral contamination and the importance of hand washing within a person's daily activities, by seeing how far the charcoal and oil is spread, and how much they accumulate in a bowl used for handwashing.

- Select one person with clean hands. Rub chilli or garlic on their fingers, and ask if they still look clean. Ask the person to lick their finger. Does it look clean, but is there something which could hurt their eyes if they rubbed them?

- Take the daily activity chart if done, and/or get people to identify when disease may be spread from dirty hands. If possible include activities such as beer-making, and giving water to small babies.

DISCUSSION POINTS

- When does transmission start?
- When does transmission occur?
- How does transmission occur?
- How can hand washing help or hinder the spread of germs that cause diarrhoea?
- What should be done to avoid this spreading?
- Why is soap liked and not liked?

NOTE TO FACILITATOR:

It is not necessary that every activity or route is identified, as long as the message has been accepted that dirt, whether seen or not seen, is passed on in many ways. The best way to stop this is where transmission begins, where faeces are left lying around, and where hands are not washed after anal cleansing.

AN EXAMPLE FROM THE FIELD

In Liyoo village, Western Province, Zambia, the water source was found to be contaminated. Yet the village had protected the source with lining, an apron and a cover. They concluded that bacteria could only be entering as a result of dirty hands on the bucket.

They installed a hand washing device with soap, and the headman required all users to wash their hands before drawing water. From that time the source has been clean, but many argue that if each person washed after going into the bush, this would not be necessary, and would stop the spread that happens in other ways as well.

ALTERNATIVE METHOD

TARGET GROUP: School children

METHOD

- Take a group of about six children from the class.
- Ask them all to wash their hands well in a bowl of water. Do they feel their hands are clean? If not, ask them to do it again. Ask others to see if they agree.
- When they are sure their hands are clean, get them to wash their hands again in a bowl of clean water.
- Then ask one of them to drink the water.
- If he or she is reluctant, ask them why.
- Summarise the discussion.

DISCUSSION POINTS

- What sort of dirt do they think is on their hands?
- How did it get there and how is it best removed?
- Can hands still be dirty when they look clean?

HAND WASHING AND SOAP

- Hand washing can be effective with less than a cup-full of water, especially if it is poured.
- Using soap removes or kills bacteria much more than washing with water alone, but using soap is usually chosen for its clean smell and moisturising properties, not its health benefit. So promote both - the effect will be equally healthy!

RECORD KEEPING
Note down the main points that come out and the degree of acceptance that dirt may be harmful but not visible. What ways of reducing risk do they see as most important?

NEXT STEPS	
To get a more detailed view of present practices and to explore the need for changing methods and times of hand-washing	**Go to Section 2.4**
To look at options for improved hand-washing and hygiene	**Go to Section 4.5-4.7**

2.4 DO HANDWASHING PRACTICES CAUSE ANY PROBLEMS?

INTRODUCTION People tend to think little about how they wash their hands, and tend only to wash them when they smell or see dirt on them. Module 2.3 showed that dirt may not always be seen, but still be very real. This exercise establishes to what degree people at present are likely to be effective in keeping their hands clean.

It can replace or be done after Module 2.3, to reinforce messages and provide a basis for planning solutions. It uses the same process as Module 2.2 on sanitation with **pocket charts** and voting.

PURPOSE This is an exercise to get people discussing what they do and whether they feel this is good practice or not.

TIME 2-3 hours

MATERIALS
Posters of handwashing (H21-26) methods and daily activities (Include G1-G4, G12, and one for defecation from S1 –S17.)

METHOD.

1. There are two stages, the first looking at **how** people wash their hands, and the second looking at **when** they do so.

2. Posters illustrating hand washing methods (H21 to H26) are explained, put in transparent folders and set out away from where people are sitting so that people cannot be seen when they are voting. Get people to select the times of day they wash hands, using the activity posters.   Make a matrix of ways of washing hands against times (see Fig 2.3), explaining it to participants and ask people to place votes according to what they did the previous day.

3. The votes can then be counted publicly and the results discussed without the embarrassment of identifying the practices of individuals.

DISCUSSION POINTS.
(see also previous Session 2.3)

- Why do people wash hands at some times rather than others?   Is this pattern good in the light of discussions on disease transmission?

- What stops people washing their hands more often?

- Are people embarrassed to be seen washing hands after defecation?

- Why do people use soap at some times and not others?

- What alternatives are there to using soap?

- Why do people use soap at some times and not others?

- Why does pouring water make a difference?

- What should be regarded as the best way or ways of washing hands and why?

NOTE TO FACILITATOR:

If there are sufficient participants (more than 15), give men and women different coloured, or shaped stones or seeds. Then the votes can be counted to see whether the practices of men and women differ. (Children can also be involved if they are old enough)

Each voter will need more than one counter for the times of day they wash their hands. Two boxes with pictures to differentiate counters for men and women may be put near the pocket charts and people select the number of counters relevant to them.

Explain that the purpose is to see, in general, what people do at present, and they should vote according to what they did the previous day.

Example of results	**From base-line surveys ....**
Practices vary widely from area to area. In Kawanda village, Mbala, men were three times more likely to wash the soil off their hands when returning from the fields, than they were to wash their hands after going to the toilet. Both men and women washed with soap quite commonly when they first got up in the morning, but did not usually use soap at any other time of day.	• In most of the developing world, families share the same bowl of water to wash their hands, and most do not use soap. Children, being most junior, wash last. • Less than one in ten children wash their hands with soap, but they are the most vulnerable, and usually least clean - members of the family.

SOME OTHER APPLICATIONS

- The information can also be used as base-line data to compare changes over time.

- The method can also be used to establish:

  ✓ Where people defecate or how they dispose of their rubbish

  ✓ What sources they use to obtain water for different purposes (previously this has also been done in schools to get a rapid picture of the 'coverage' and problems over a large area)

  ✓ Practices in water collection and storage

  ✓ Identifying the existing and desired level of sanitation or water supply, using 'ladders' depicting different levels of technology (see Sections 4.1-4.4)

RECORD KEEPING	**NEXT STEPS.**
You should make a note of the matrix made, including the results of how many men, women and children vote in each square (category). Also note down special reasons people give for **why** they do what they say they do.	Go to Modules 4.5-4.7 which explore the options for improved hygiene and water use at the source and at household level. Modules 4.8 and 4.9 look at personal hygiene.

Table 2.3 : EXAMPLE OF A MATRIX: TIMES OF HANDWASHING

(42 participants : 21 women, 8 men, 13 children)

HOW / WHEN	No soap	With soap	No soap	With soap	No soap	With soap
On getting up:	On getting up: 6 women 2 men 2 children	3 women 1 man 1 child				1 man
After latrine:	After latrine: 2 women 1 man	1 woman		1 woman 1 man		
Before — Before eating:	Before eating: 4 women, 4 men 6 children	3 women	Before: 2 women 4 men 4 children		Before: 1 man 3 children	1 man
After — After eating:	After eating: 10 women 2 men 6 children	After: 3 women 4 men 1 child		After: 2 women 1 man 2 children		After: 1 man
Before cooking:	Before cooking: 2 women		1 woman			
After cleaning baby:	After cleaning baby: 2 women	1 woman				

Fig 2.3   HANDWASHING METHODS

2.5 DO WE HAVE ENOUGH WATER IN THE HOUSE?

INTRODUCTION This aspect comes up frequently when handwashing and personal hygiene are discussed as in 2.2 and 2.3 but what are the underlying reasons for this?

PURPOSE

To isolate different factors which lead to insufficient water for good hygiene.

MATERIALS

Posters of daily activities (**G1-12, H21, 23, 25)** , seeds/ stones,

TIME 2 hours

METHOD

1. Divide men, women (and children, if many) into separate groups. Using **daily activity analysis** get each group to identify the different activities which they carry out from the time they get up, which relate to collecting and using water, going to the latrine or bush and washing children, hands, clothes, dishes, and themselves. Posters may be used, or pictures drawn. Allow more than one copy of activities such as water collecting, and encourage people to draw additional pictures where they feel they are needed.

2. Ask people to put a stone/ seed on those activities/times of day when they usually have less water than they would like or a problem relating to this.

DISCUSSION POINTS

* Do men/ women generally feel they have sufficient water for their needs?

* **What are the main constraints to more water use in the house?**

  These may include -:

  ✓ Distance to source?
  ✓ Capacity of containers?
  ✓ Amount of water at the source/ numbers of users?
  ✓ Lack of assistance by other family members than the women?
  ✓ Lack of privacy to bathe?

  ✓ Too many other tasks for women?
  ✓ Fear of source drying up?

* **Why do these constraints exist?**
  ✓ Lack of previous discussion?
  ✓ Lack of awareness of problem by those who could help?
  ✓ Lack of funds for more containers/ more soap?
  ✓ Too few sources?

Example of results

In Northern Province people like to store water overnight, but almost two-thirds of those asked felt they could not store water long enough because of limited container capacity. This meant more frequent trips to the source. Others found that they had difficulty to carry enough water for all the washing needs of the family.

From baseline and other surveys

* Of 1150 sources, 58% always had enough water for user demand, a third had seasonal shortages and just under 10% never had enough water.

* Surveys show that in the last ten years, men and boys have increased the amount of water they carry, girls' contribution remains the same, but women's water carriage has reduced by 25% - men CAN and DO carry water and can be encouraged that they are not unusual in this!! (see Section 4.5)

* Household water container capacity tends to affect water consumption more than the distance to the source does.

Fig 2.4 POSTERS: TIMES/ ACTIVITIES - ADD MORE!

G.1 Day break, getting up

G.2 Eating a meal

G.3 Cooking /preparing food

G.4 Working in the fields

G.5 Cleaning up around the house

G.6 Beer drinker

G.7 Night-time

G.8 Washing clothes

G.9 Washing dishes

G.10 Collecting wood

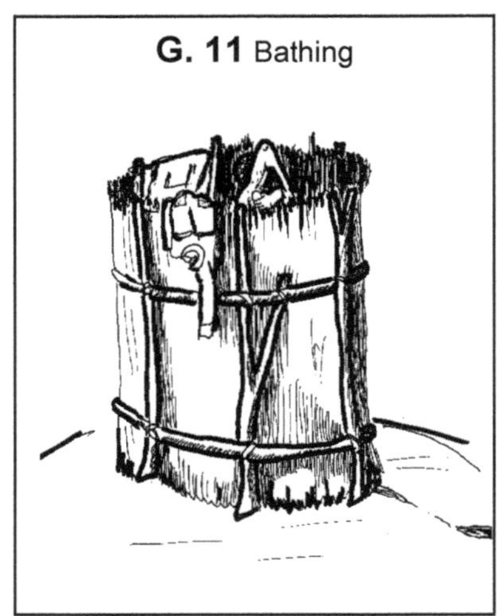

G. 11 Bathing

G.12 Cleaning up the baby.

G.13 Boiling water

G.14 Visiting the clinic

W.25 Collecting water

SECTION THREE

43

BUILDING COMMUNITY CAPACITY TO CHANGE THE SITUATION

GENERAL TOPICS

3.1   BUILDING ON EXPERIENCE

3.2   STAGES IN UNDERTAKING A PROJECT

3.3   ARE THERE ANY RESOURCES AVAILABLE?

3.4   CAN IMPROVED WATER SUPPLY PAY FOR ITSELF?

3.5   ACHIEVING GOALS A STEP AT A TIME

3.6   NEIGHBOURS AND FRIENDS HAVE GOOD IDEAS

3. BUILDING COMMUNITY CAPACITY TO CHANGE THE SITUATION

INTRODUCTION

Most communities/ households initially believe that they have few resources, skills and little capacity to solve the problems they have identified and which they regard as priorities for change. If that were not so, they would have taken some action already.

Where a community comes to you for assistance having already identified the problem they want to solve, this is the stage at which you would start to work with them.

The role of the facilitator is to help them to see their strengths and potential to change the situation, and then to provide the necessary advice and technical assistance or show the community where to obtain it. They have skills and materials which may be used in construction and improvement. Alternatively these skills and materials can be used to raise funds and these funds can be used to purchase the materials and skills they do not have, or to make the necessary contributions to an investment fund so that they qualify for more assistance.

One important point is that improving the water supply may cost money and need labour, but much more could be done to make the water source also provide income to the users and so not just lead to benefits in health (Modules 1.3 and 3.4).

This section shows the application of some tools to enable communities to feel that they CAN succeed and also to identify how big a step they are able to take at one time.

From that point they can discuss what options they have for solving their problems, with the resources they have or can call upon, and then make an action plan.

3.1   BUILDING ON EXPERIENCE.

INTRODUCTION   Exploring the past history of the community can help identify the things that people are proud of, and the successes and failures which can provide lessons which assist in future action

PURPOSE To identify what aspects of the community are regarded with pride, the management and other skills available, groupings within the community and the standards of behaviour that have existed and how they were encouraged/ enforced.

TIME. 2 hours

MATERIALS. Paper and pen/ pencil

METHOD

1. Draw a line on the ground representing the time over which people have memories of the history of the village. This **historical timeline** will have one end representing the oldest remembered event and the other being the present.

2. Ask older members of the community to identify specific events in between which affected the village, and mark these on the line.

3. Summarise the main ideas that come out in discussion, which can be built upon and what can be learnt from them.

4. Transfer the line and main events to paper and decide who will keep it, so that it, like the community map, can be used again for discussions of other issues.

DISCUSSION POINTS

- Identify the achievements and events of which people are proud, and the aspects in which the community failed to achieve what they wanted to do. What were contributing reasons? Would the same happen now?

- Ask the older members to describe how the village was in their childhood, and what were the way of life and the rules of behaviour in relation to water, hygiene and sanitation.

- Are the changes in beliefs, attitudes and practices an improvement or not? Why? Is there a difference between the views of the older and younger members of the community?

- Ask people also to discuss how decisions were made then and now, whose ideas are most listened to. Who were or are key personalities, and what was so special about them?

- Is the community composed of different groups, and how does this affect management and decision-making?

- How can all groups (economic, tribal, age, ability) be included in any plans and decisions? Are there people who are often excluded or not considered when decisions are made?

NOTE TO FACILITATOR:

Ensure that there is a good cross-section of age groups present. Make sure the views of elders and younger members are both heard.

Summarise the lessons learnt, and emphasise aspects of the history of which the community should be proud.

Identify with them the features of successful management and decision-making that contributed to successes, (which may have been at family as well as community level) and problems which were or were not overcome.

Some other applications	**NEXT STEPS**	

Some other applications

Historical timelines can also be used to discuss changes in:

- Diet and nutrition
- Health, economic or educational status
- Pressure on resources

NEXT STEPS	
If present organisation seems to be weak	**Go to Module 3.2**
If community feel they cannot afford change	**Go to Module 3.3-3.5**
If community wants change, but does not feel capable	**Go to Module 3.2-3.5**

RECORD KEEPING
Note down the main features of the community which came out. Include who are the most influential people and what groups tend to be marginalized (excluded or not considered). Does the community tend to act as one, through a strong headman, or is it composed of several equally strong groups who do not tend to agree on the best way forward. Note any examples of how the community has previously overcome problems.

3.2 STAGES IN UNDERTAKING A PROJECT

INTRODUCTION This introduces the idea of a project cycle, so that people think of the stages they will need to go through to reach their goal.

PURPOSE

To explain the project cycle to a community in a form that they can understand and use in their setting. It prepares them for planning an intervention so that they consider all aspects.

MATERIALS Pen and paper

TARGET GROUP. This can be done with just the water and sanitation / development committee or with a wider cross-section of the community as a whole.

METHOD

Story with a gap. This starts with a man and his family sitting thinking of building a new house on an empty plot and ends with a poster of a well built house.

The discussions which fill in between these two pictures outline the activities which need to take place.

Notes are taken of the main activities identified and then fitted into project stages.

TIME 2 hours

DISCUSSION POINTS

Stages should include not just the materials needed and the order in which the house is constructed, but may also include aspects such as:

- Reason or felt need for someone to having own accommodation

- Identification of site for building a house

- Consultation with relevant authorities e.g. village headman, chief, family members

- Consultation with extension staff e.g. extension worker

- Assessment of ability to build a house – resource availability that will include building materials, building tools, skilled and unskilled human resource

- How to ensure co-operation from others

- Determination of size and shape of the house and inclusion of house accessories such as kitchen, latrine, bathing shelter, dish rack, rubbish pit, food storage and processing area (+planning and design)

- Scheduling of activities, task allocation and materials needed for each step

- Progress monitoring at each stage ensuring the house construction process is satisfactory from foundation, roofing completion of the house and other accessories

- Evaluating what things might have been done differently/ more cheaply and what others changes could still be made.

NOTE TO FACILITATOR:

Discuss with participants the effect of any of these activities being missed out.

Note with participants which of the activities identified are relevant to the changes they are planning to make, and use this as a framework for action planning. (See Section 5.)

NEXT STEPS

Keep notes from this session and remind people of the results of discussion when using Module 5.1

Continue with Section 3, Modules 3-3.5.

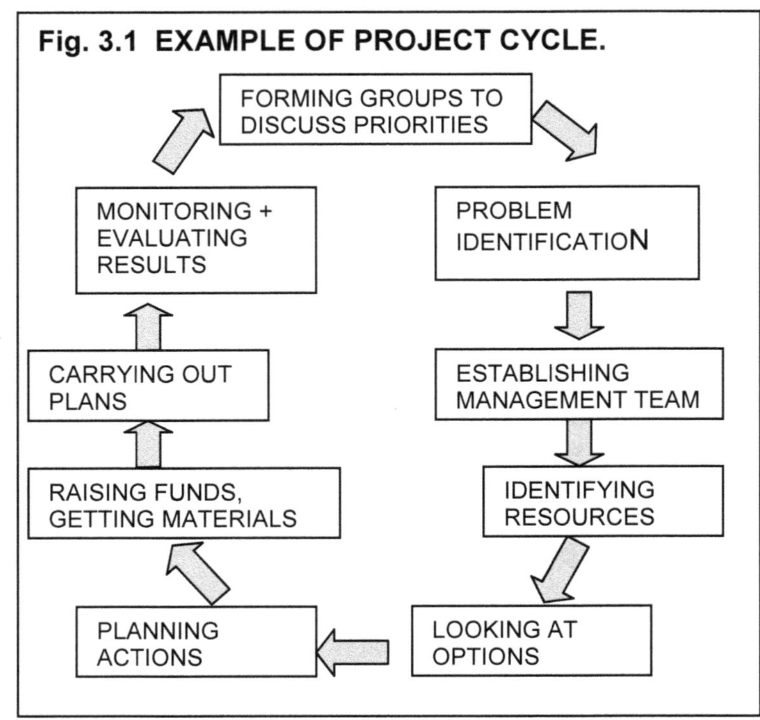

Other applications

This discussion is relevant to almost any area of community development. It helps people to put activities in an order that ensures that funds are ready before materials or skills need to be paid for, and that materials are at the site before construction requires them. It also makes sure, among other things, that users are consulted and sources of skilled advice identified before construction begins, and may emphasise the need for clear allocation of responsibilities by the community.

Fig 3.2 POSTER EXAMPLES: HOUSE BUILDING - make ones relevant to local house types and standards

The plot	Building a big house	Big house completed

RECORD KEEPING

Note down the main steps that people felt it was necessary to follow when considering a project and the activities/responsibilities.

This might include

- Identifying what changes are needed
- Defining how decisions will be taken and who has the last word
- Designing the new facility or changes to be made including
    - Materials needed
    - Labour and skills required
    - Considering factors which may affect the design
- Planning the work
    - Dividing the work into stages if necessary
    - Times and seasons for completing different stages
    - Giving people specific responsibilities within the team
- Judging progress and bottlenecks
- Looking back at how things could have been done differently and what has been achieved (see also Sections 6.2-6.4)

3.3 ARE THERE ANY RESOURCES AVAILABLE?

INTRODUCTION Rural communities tend to be regarded, both by themselves and by others as having no resources or relevant skills. They say they are too poor to buy materials, and others should come and assist them. Certainly a part of the facilitator's role is to see what assistance may be available, but it is also to help the community see for itself that it is capable of making changes which can have a big effect and which are affordable and sustainable.

Fig 3.3 Seasonality calendar discussion at Liyoo village, Sefula

PURPOSE

This session helps communities to see what skills and materials they may have to work with, and also where there is scope for raising additional cash/ kind over the year.

It applies particularly where the changes wanted by the community require improvement to existing, or new, facilities such as water supply or latrines.

TIME 2-3 hours

MATERIALS   Community map from Module 1.2. Pen and 12 pieces of paper (or mark out months on ground).

METHOD

1. Use the existing community map or if none, construct a map (see Section 1.2), and then draw up a **seasonality calendar**.

2. Firstly, using the **community m ap,** 'tour' the village with participants. Going from household to household (while sitting round the map), identify the skills available in each, and any interesting ways local materials have been used in construction (or re-cap if this has been done previously).

3. Add to the map any income generating activities not already drawn in (eg. beer making, mat weaving, bee keeping) noting the skills and the local materials used.

4. **Seasonality calendar.** On the ground, or on sheets of A4 paper set out an area (or page) representing each month of the year. Note what crops are harvested and what other activities bring income in each month.

Some other applications of seasonal calendars:

- Identifying disease patterns and planning preventive campaigns

- Improving food security

- Planning to undertake source improvements when water levels are lowest and funds available.

DISCUSSION POINTS

- What crops are produced in sufficient quantity to be sold for cash or exchanged for kind?

- At what times of year would fund-raising be most likely to succeed?

- What local skills can help generate income for fund-raising?

- What local materials can be used in construction, and when are they available?

- At what times of year are people most able to provide skills/ labour for a project?

- Ask people to make two groups of resources and skills– those which can be used directly, those which can be produced in surplus and sold for payment in kind or cash

- Keep these lists and the map for use again with Module 5.1 for planning who will do what and when, to implement changes.

RESULTS OF COMMUNITY SURVEYS IN ZAMBIA

The mix of sources of income was found to be varied, with only just over half of all households depending solely on agriculture for their income. Many families trade in small commodities, agricultural produce, beer, locally made mats etc.

Almost 40% of households sold some of their harvest, even though a quarter of these households were then short of food later in the year.

RESULTS OF COMMUNITY SURVEYS IN ZAMBIA

- Over three quarters of communities had at least one brick-layer and most had someone who could repair bicycles. All had a carpenter / woodworker.

- Local bricks can easily be used to line a well, or create a well head on which a windlass or pump can be mounted, and good local brick-layers can quickly be trained in brick-lining for wells and latrines.

NEXT STEPS	
If community feels resources are not sufficient	**Go to Module 3.3-3.5.**
If communities feel they are ready to define better the changes they have the capacity to make	**Go to Section 4. (You can come back to 3.4 / 3.5 afterwards if necessary.)**

3.4 CAN IMPROVED WATER SUPPLY PAY FOR ITSELF?
(See also Module 1.3.)

INTRODUCTION.

Initially it may seem that people are having to pay to have better water, but, especially where there are few users (family wells) it is usually possible for owners/users to grow vegetables all year or brew beer and so to avoid having to take cash from their household budget. In fact most make a profit which means they can make further improvements, such as installing a device to lift the water more easily.

Earning money from the source should be considered from the start. Then people can see their supply as an asset (a thing of value to them) which can cover its own costs and be up-graded over time. This change in attitude may encourage people to make improvements, or dig more wells and include aspects which will make income generation easier, when they are making plans.

This session uses Focal Group Discussion and an Open-Ended Story, along with the Seasonal Calendar developed in the last module. If there is an Agricultural Block Extension Officer, ask them if they can assist you.

Fig 3.4 Tank for watering plants: A Kaoma farmer made a tank so that watering plants is easier. He can draw all the water and spread it by hose pipe to plants.

PURPOSE To explore whether there could be ways of making money from the water source and seeing what changes would be needed to make these easier.

TIME 1.5 hours

MATERIALS
Seasonal calendar from 3.3.

METHOD

1. If there are several family wells within one community, ensure that as many men as possible attend the meeting as well as women, and that well owners are present.

2. Go through the seasonal calendar made in the previous module (3.3). Ask participants to discuss in two (mixed) groups -:

   a) where yields could be improved if seedlings could have been raised before the rains came and

   b) what crops could be grown with small-scale irrigation in the dry season.

   c) availability of suitable plots near the water sources.

3. If there appear to be possibilities and interest, tell an open-ended story to start discussion on what ways water use could be developed to bring in money.

DISCUSSION POINTS.

- In what ways can water be used to increase income a) if land for cultivation is available and b) if it is not?

- How can watering plants and drawing water be made easier?

- Can drainage water be used rather than just soak away? (For example sugar cane or bananas can be planted at the end of the drainage channel. See Poster W 42.)

- In what ways would the well and water lifting need to be changed if more water was going to be used?

- Would people be able to excavate new wells more often if they could recover the cost of digging and materials after the well was made? Could a credit system be set up?

- What other considerations are there in irrigating from the well? (for example depth of water/ reliability, drainage water ponding, labour, market for produce)

NOTE TO FACILITATOR:

1. In summing up, also refer back to the conclusions of Module 1.3, which drew attention to the savings that can be made when diarrhoeal diseases are reduced. This can also be done where sanitation rather than water supply is the focus of interest.

2. Similar cultivation has been done at communal wells to raise funds for spare parts. However in this case land-ownership and who gets the income can cause problems. Try to find examples in your area which may be familiar to participants, and will show them what has been done elsewhere.

3. Note that water quality usually improves in sources where more water is taken out.

Improving the efficiency of water lifting

A litre of water weighs one kilo. Lifting water to irrigate crops can mean lifting several hundred kilos a day. Using a windlass or pulley takes the weight off the person drawing water and a handpump increases the amount of water which can be lifted for the same effort. These are systems which could help increase crop yields and so pay for themselves over quite short periods - windlass within a year; hand or treadle pump within three years.

Improving efficiency of watering

Drawing water and taking it to plants is often difficult, especially if there is only one person working. Water can only be drawn when empty containers are available, and then the farmer must take each container around the plot. If there is a cistern, he can first draw all the water and then take it to the plants. With a hose pipe from the cistern or tank, he can direct the water to plants without spillage, loss of water in a channel, or the effort of carrying every litre of water to the place it is needed.

OPEN-ENDED STORY

Mr Kalunga had a deep lined well in a plot where he grew tobacco and maize. His tobacco crop was often small, since the plants were not fully developed when the ground dried out after the rains. His wife also complained at being tired and having back-ache because the water level was so deep for drawing water. He saved up and bought an iron bar, which the blacksmith cut and bent for him to made handles. He made the barrel for a windlass from mubanga wood and set it up over the well.

His wife was very happy and that year he grew seedlings before the rains came and in the dry season he grew tomatoes and gourds which he sold, but he found that carrying the water to each plant was hard work. With the extra cash he bought an oil drum and he exchanged 2 gourds for a length of hose pipe. The next year......

Discuss how else the water could be used and what other changes he could make with the extra cash so that life improves.

NEXT STEPS	
If resources are limited to plan for income generation	**Go to 3.5.**
If not	**Go direct to 4.1 / 4.2.**

RECORD KEEPING

Note down if people are interested to follow up on the idea of making income from their water supply. If so it may affect the technologies they will be most interested in and some of the features they should plan for even if they do not install them from the start. (See Modules 4.1 and 4.2 and 5.1.)

3.5 ACHIEVING GOALS A STEP AT A TIME.

INTRODUCTION

This explores the idea that one may be able to achieve a higher standard of facility, or greater change to way of life, by identifying intermediate steps, than by trying to complete the whole project at one time. Although one may wish for a house of brick with glass in the windows, or a lined well with a windlass and apron, this may only be achievable if done a bit at a time. So each year further improvements are made as funds are available, until the final product is achieved. However the early plans include how to reach the final stage, so that each step allows for future up-grading.

Fig 3.5 Changing a house step by step

Can this house be turned, step by step ……

....into this house?

MATERIALS   Sample Posters G18, G17, A4 paper and pens

TIME 2 hours

METHOD   **Story with a gap 2**   Ask participants to imagine that they have a small amount of surplus produce each year to sell, (say equivalent to 3 pockets of cement) and they want to build a good house. Discuss whether they could afford to build the fine brick house with zinc roof, glass windows, concrete floor and metal lintel with a fine door, or would they have to make a house out of more local materials?

Would it be easier to build such a fine house, if it was done a bit at a time?

Get someone to draw the changes as they are agreed and to make a ladder of them.

DISCUSSION POINTS

- In order to change from one type of house to the other, while living in it, what would you change first? (eg. would you change first from grass to brick walls, or from grass to zinc roof). Would you make a hole for a window before you could afford the glass?

- Discuss whether in the same way, the changes they would like to make to their water supply, community sanitation strategy, or hygiene behaviour could also be done bit by bit if resources are limited.

- Will people lose interest if the project is not completed in one step?

NOTE TO FACILITATOR:

If water supply improvement is specifically being considered, the exercise can be replaced by looking at ladders/ options and seeing whether the final product desired can be achieved at one step with available resources, or whether it should be phased, as with house up-grading. (see Section 5.) This would form the first stage of planning for solutions.

Other applications

The same discussions may be held in preparation for considering technical options for latrines.

The same concepts are relevant to consideration of other construction programmes, but also in phasing activities within:

- Community sanitation strategies

- Community development programmes

- Changing behavioural norms, especially among larger groups (community rather than family)

AN EXAMPLE FROM ZAMBIA

Liyoo village near Sefula, Western Province, had a scoophole that was often smelly from run-off and leaves falling in. The community decided to raise funds and buy cement to make rings and line the source. They also made a top slab. The following year they collected for cement for an apron, and finally someone produced an old drain cover as a lid and they bought a communal bucket. After four years, they considered putting on a windlass, but the hole in the top slab was too small and they felt it was not worth breaking it and making a new one, as they would also have had to destroy the apron to make sockets for the poles. They had got further than if they had tried to do it all at once, but realised that there are some things they would have done differently, if they had thought of how they would like the final product to be, from the start.

Fig 3.6 Step by step improvements of wells, scoopholes and latrines

Family well

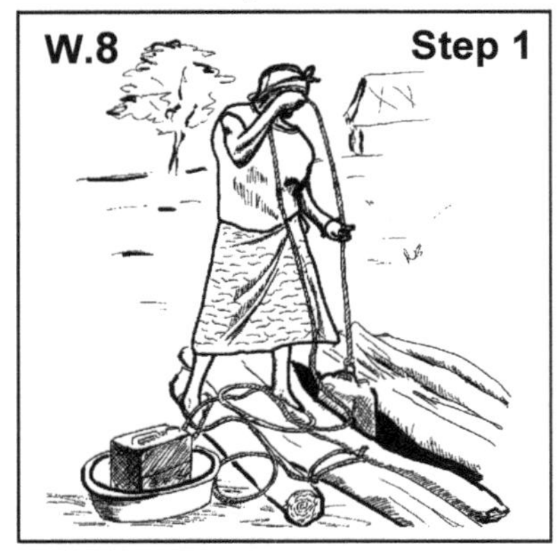

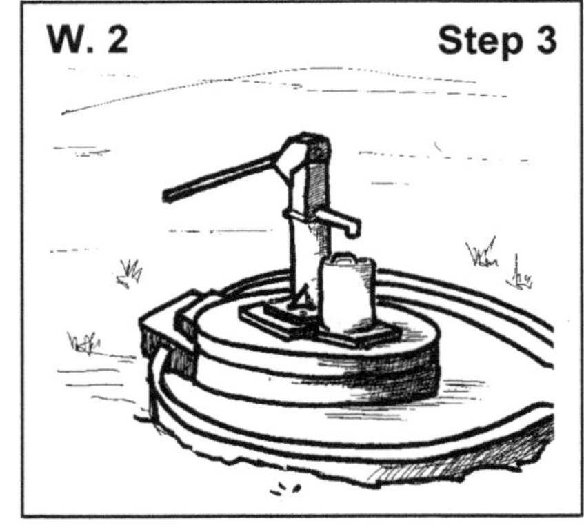

Scoophole

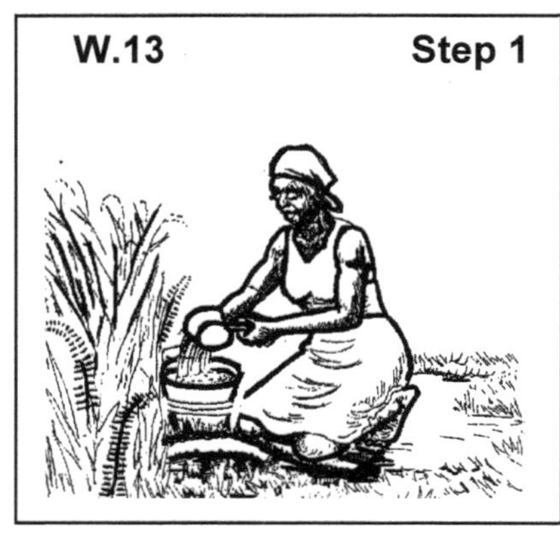

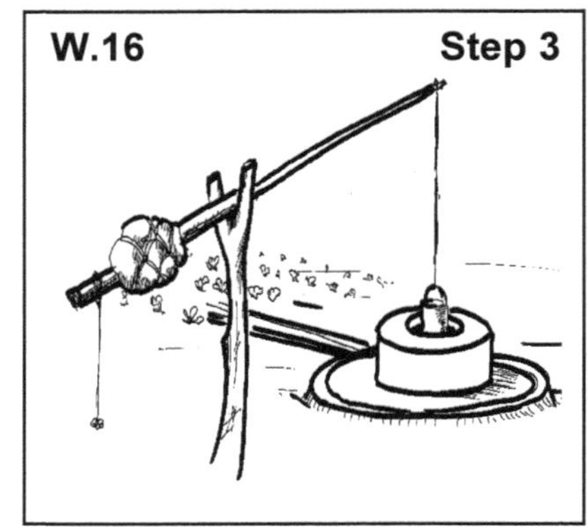

Latrine

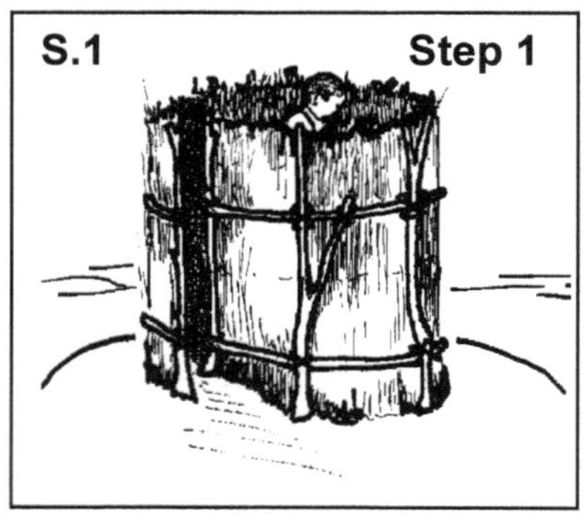

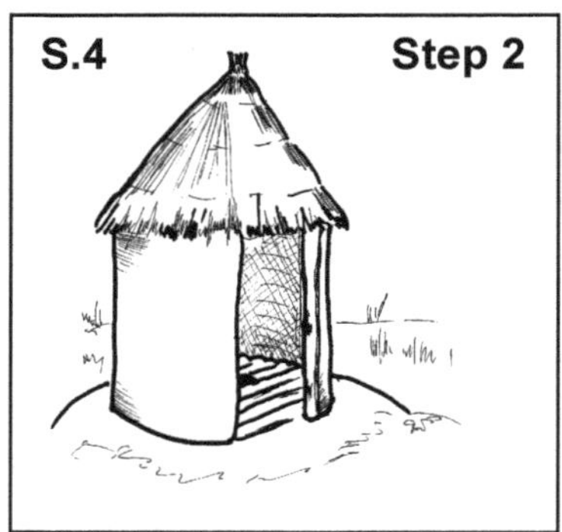

3.6 NEIGHBOURS AND FRIENDS MAY HAVE GOOD IDEAS.

(Organising An Exchange Visit.)

INTRODUCTION

An exchange visit consists of facilitating a small group of community members to pay a visit to a community that has dwelt with similar problems. It is a powerful tool that provides an opportunity for community members to take a structured look at issues important to them in another community. It is also an opportunity to look at how other communities have solved problems similar to their own. They can get a broad view and possible solutions through suggestions from the host community. It increases communication within and among communities.

Fig 3.7 Sailunga: Men from both communities got together with facilitators for a discussion, and then joined with the women.

PURPOSE

To exchange and share ideas and experiences related to practical and management issues and to provide an opportunity for participants to move away from their own situation and reflect from their problems as seen in a mirror. It alerts them to aspects they may not have considered, and problems which others may have found difficulty to overcome as well as to solutions.

TIME One day

MATERIALS Check-list of issues which they wish to raise

METHOD

1. Ask concerned community / institution to consider the idea of visiting counterparts in another community.

2. Establish rapport with the community to be visited and get consent.

3. Plan the date, purpose and agenda for the visit.

4. Allow people to select their own representatives to take part in the exchange visit.

5. Arrange a meeting prior to undertaking the study tour. Allow the community members to agree on the focus of the visit and prepare a check-list with points for discussion and agree on Do's and Don'ts, while visiting the community.

6. Inform the host community about when and why the visit will take place, and ask them to think of issues they would mention in relation to the topic selected by the visitors.

7. At the time of the visit ensure that the host community talks about the historical background of the interventions, successes and constraints.

8. During the wrap-up session, ensure that both host and visiting communities share what they have learnt from the other party and what type of advice they would like to receive or give one another.

NOTE TO THE FACILITATOR:

You may need to make special arrangements for women taking part in the exchange visit, especially if they are staying overnight.

You will need to draw up a check list with a few members of the visiting group, before you go.

NEXT STEP

After the exchange visit, it is important that new ideas are fed back to the whole community. A discussion should take place about how the new ideas can be incorporated in each community and how this might affect plans.

RECORD KEEPING

Note down the issues raised and the experience which may prove helpful when the community is planning changes. Make sure you include not just practical ones, but those relating to management issues and overcoming social problems, as well as technical ones.

SECTION FOUR

LOOKING FOR SOLUTIONS, SELECTING OPTIONS

4.1 WATER SUPPLY TECHNOLOGIES

4.2 IMPROVING EXISTING SOURCES

4.3 SANITATION

4.4 BEHAVIOURAL STANDARDS AT THE SOURCE

4.5 INCREASING AVAILABILITY OF WATER IN THE HOUSE

4.6 HANDWASHING

4.7 HYGIENE AROUND THE HOUSE

4.8 IMPROVING PERSONAL HYGIENE (1)

4.9 IMPROVING PERSONAL HYGIENE (2)

4. LOOKING FOR SOLUTIONS, SELECTING OPTIONS

INTRODUCTION

This section examines the choices available to solve the problems which have been i dentified in previous sections, and gives an opportunity for the selection of particular options. Where the community needs to plan how these options will be implemented, this is discussed through modules in Section 5.

Technology options for water supply may be tackled in two different ways. You may start with getting users to see how they could improve their own source (Module 4.2) and if this is not satisfactory, then move to considering other, more expensive options (Module 4.1). This would be the best approach for smaller, poorer communities (especially where less than 20 hous eholds.) For larger ones, the ladder of alternatives (Module 4.1) could be used first, and i f people feel they prefer their own source, or that they cannot afford the long term commitment to more expensive sources, then continue to Module 4.2.

The role of the facilitator is to describe unfamiliar options which are relevant/available, and to provide information chiefly in response to questions from participants. Some background information is given and reference should also be made to any theoretical guidelines provided by relevant ministries (Health, Local Government, Social Development, Water Affairs.)

 At the end o f the modules in this section, the changes that a community intend to make, whether behavioural or physical, should be clearly defined, so that they are then ready to plan and implement specific improvements and d raw up any proposals for assistance, based on the needs they have identified.

IMPORTANT NOTE

Research in Zambia has shown that contrary to popular belief, protected water sources, particularly lined wells with aprons, drainage and windlasses, as constructed by the Department of Water Affairs and others, do not necessarily provide a safer, more reliable supply than traditional sources, especially ones with some protection. They are no more likely to have safe water than a scoophole in a sandy aquifer, and are generally only 10% more reliable than family (unlined) wells.

Handpumps and boreholes do usually provide a slightly better quality water, but this has to be balanced against the risk they will stop working. Availability of spare parts and their real cost, is still a major problem for users over much of the country, so that often the ability to provide the highest quality water is also associated with a supply which may not be functioning.

These basic facts should be taken into consideration when discussing change with communities and their (and your) experience of the performance of different systems should be given more weight in discussions than any theoretical benefits of different systems. Few systems perform according to their ideal design. Small changes to a source can bring big improvements to water quality and almost always lead to less than 10 faecal coliform/100ml. See also Table 4.1.

4.1 TECHNOLOGY OPTIONS FOR RURAL WATER SUPPLY

INTRODUCTION This session explores the different ways in which water from the ground can be br ought out most safely and eas ily. It requires people to think about many different aspects and will draw on their experience both of the existing source and any others they have used while staying in other villages. It requires the facilitator to be abl e to answer questions from the group, and for this purpose background information is attached. It uses simple or pair-wise ranking to establish the most popular options and the reasons for their choice.

PURPOSE To allow the community to have adequate information to make a well-informed decision on the technology they would like to adopt.

TIME 2-3 hours

MATERIALS
A4 paper, pens, flip chart paper and posters of alternatives.

TARGET G ROUP Men and w omen, preferably discussing in separate groups.

METHOD

1. The aim is first to define what alternatives there are, then to discuss the advantages and disadvantages of each, and t hen to make a ' ladder' ranking them according to preference.

2. Firstly, ask all those who have lived in other villages to draw (or pick from the pile of pictures if available) and explain the systems they have used and what they liked and did not like about them. This may include a drawing representing their existing source (from Module 2.1).

3. Ask them to explain what else is known of how these systems have worked in nearby communities, especially how the community has kept the system working, or why they couldn't manage to do so.

4. What additional information would people like to help them make their decision?

5. Summarise the findings from their experience and add additional information if desired.

6. Ask women and men separately (women first if possible) to put the posters in the order of their preferences and ex plain why each is placed where it is. If the order is not the same, continue discussion until agreement is reached on w hich option(s) will be followed up in the next session.

DISCUSSION P OINTS - for each alternative

*   **Reliability** How reliable are similar systems in neighbouring villages? Do they go dry? What other reasons are there why the supply stops working?

*   **Accessibility** How easy are they to use, for domestic and irrigation or other purposes?

*   **Water qual ity** Does the water taste good, and i s it regarded as clean by users?

*   **Cost.** What is the likely cost and how much of this is the community likely to have to pay, both for implementation and as a r egular cost for maintenance? What work would be involved and how much of it would the community need to do?

*   **Funds.** Is the community able to raise these funds or will they need to make a request?

*   **Preference.** What are the reasons for preferring one t o another? What features does each have which are regarded as 'good', and what benefits would each bring?

- **Skills.** Which options can the community undertake without additional advice or expertise, and for which would they need training or supervisory assistance?

- **Use of cont ractors.** Which options or parts of the work could be undertaken by contractors and which by users themselves? A re contractors available?

NOTES FOR FACILITATOR:

Before this session make sure you have the following information:

- Availability of funds from district for various technology options, and av ailability of contractors/ well diggers locally

- Any NGO activities which might support rural water supply development. (Should be through district anyway, but may not be.)

- Check with the community whether they are prepared to consider improving their water supply even if little or no f unds are available.

> **Relevant posters   W1-15**

Background information for the facilitator:

This is given to assist in answering any questions participants may have. I t is suggested that the information is **NOT** used before groups or individuals have given their own experiences and ideas, or have asked for specific information. If additional information is required consult with the Department of Water Affairs and the Central Board of Health.

It is important that users are aware of the construction costs and the running costs for different options. This includes both cash and their commitment in labour and provision of materials, to ensure long-term operation of facilities.

To convert dollars to approximate local currency multiply by the ruling rate at the time. In February 2002 the ruling rate in Zambia was 4,000 kwacha to the dollar, so $1000 was then four million kwacha.

How much will it cost to make?

Source

A **borehole** normally costs from $2,500 to $4,500 to drill and complete.

A **Department of Water Affairs hand-dug well**, lined with concrete rings, normally costs from $2,000 to $3,500 to construct.

Handpump costs vary with depth from around $300 to over $1,000.

A **windlass** costs around $120 including chain, if made of steel. A wooden windlass barrel with metal handles costs around $10. A pulley will cost around $10 or can be made yourself.

An **improved traditional source** will cost between $100 and $400 depending on the amount of lining and type of lifting device.

How much will it cost to keep working?

A guideline is that over ten years it will cost at least 2% per year of the cost of construction to maintain the system. In the first years the cost may be very low, until the chains or bearings wear out, or the well needs re-deepening. Costs for the first years will be around $10-20. However, it is usual to have major repair needs within ten years and funds need to be accumulated for this. It is necessary to have an accounting system and management where such large sums of money are involved.

Guideline long-term annual maintenance costs:

Handpump + borehole	$60-100
Shallow well and windlass / pulley	$50-75
Traditional source	$10-20

Table 4.1 RURAL WATER SUPPLY TECHNOLOGY OPTIONS (Background information for facilitators)

All costs are approximate and will vary between areas

SOURCE TYPE	SUITABLE CONDITIONS	ADVANTAGES	DISADVANTAGES
Improved springs and scoopholes *Cost $US 30-100* **Posters W9, W10, W12**	• Where water is found near the surface (within 1.5 metres)	• Small low cost improvements can bring good quality and greater reliability to water supply • Systems of management and ' ownership' already in place • Taste usually more acceptable • Maintenance costs minimal	• Distance to collect water not reduced • Some risk of contamination • May dry up • Yield sometimes limits number of users
Improved un-lined (family) wells *Cost US$ 30-100 (Improvement)* *Cost US$ 150-350 (new)* *Part or full lining, making parapet, top slab drainage and soakaway. May include lifting device (eg. Windlass)* **Posters W2, W3-W7**	• Well consolidated ground • Water within 15m and seasonal fluctuations less than 4m • Small scattered communities	• Convenient, short distance for water collection • Low cost and easy to construct or improve • Users often more keen to invest in privately owned supplies than communal • Rapid improvement in quality, • Uses only locally available skills and materials • Minimal maintenance costs • Easy to up-grade in stages to pumped well • Easily re-deepened • Can water crops and increase income	• Suitable only in cemented/ consolidated ground • Some risk of contamination • May dry up
Lined Communal hand-dug well with bucket & windlass or with pulley *Average cost-* *US$2,000 new* *US$ 1,500 rehabilitated* **Poster W4**	• Depth to water less than 15m • Total depth less than 25m • Water level fluctuations < 6m • Ground well-consolidated but not strongly cemented. • Advantageous where the inflow of groundwater is slow, or there is only a thin layer of freshwater overlying salty water, because of the large storage available. • May dry out quickly where the aquifer is thin or there is downward flow into another aquifer beneath it.	• Larger storage capacity able to serve larger number of people • Water visible • Familiar technology, especially to Department of Water Affairs • Possible (but difficult) to re-deepen • Community participation fosters ownership • Available water even when lifting device broken • Can easily retrieve bucket • Easy to lift water. • Drainage water can be used for communal veg. garden.	• Expensive to construct • Risk of drying • Needs much labour in construction • Often difficult to site successfully near houses • Does not always reduce distance to water • Takes long to construct • Expensive to clean • Things can fall in • Risk of contamination • Needs regular replacement of buckets and chains • Difficulties in finding competent contractors • Problems of communal maintenance & management

Table 4.1 RURAL WATER SUPPLY TECHNOLOGY OPTIONS (continued)

SOURCE TYPE	SUITABLE CONDITIONS	ADVANTAGES	DISADVANTAGES
Jetted wells *Cost US$ 1,300 including* **hand pump** *Shallow small diameter well constructed by forcing casing or pipe downwards with the pressure of water pumped through it* **Poster W1**	• Depth to water < 8m • Aeolian and alluvial sands	• Very fast to complete (3 a day possible) • Can be re-jetted if it clogs • Relatively cheap equipment for construction • Uses light equipment to construct, good for remote / difficult access areas • Sealed from contamination • Hand pump able to serve up to 250 people	• Only suitable for sandy areas • Depth limited to about 18m • Can't draw water if handpump fails • Needs pump spare parts • No contractors at present • Well screen can be expensive and may not be available locally • Communal management
Hand-augered with bucket pump *Cost US$ 1,750 inc. pump* **Poster W3**	• Depth to water less than 10m • Total depth < 25m • No great variation in ground conditions, especially no cobbles.	• Easy to construct (simple technology) • Low contamination • More reliable than shallow wells • Stronger buckets • Few moving parts to maintain • Fast to construct • User friendly • Can be done by small communities • Cheap	• Unsuccessful in many ground conditions • Not familiar technology • Water not visible • Impossible to re-deepen • Limited volumes of water, slow lifting • Difficult to retrieve buckets • Difficult to buy new ones • Risk of contamination • Few organisations have equipment
Borehole with handpump *Cost US$ 3,500-5,000 inc handpump* **Poster W1**	• Depth to water < 40m • Total depth < 100m • Water level fluctuations and aquifer type not a constraint. • Not suitable where thin freshwater lens lies over brackish water	• Safer water • Goes deeper. more reliable • Labour reduced • Less technical problems underground • Fast to complete • Can site near houses • Plenty of contractors to choose from • Easy to lift water if within 30m of surface • Can serve up to 250 people	• Expensive • Difficult to clean • Not familiar/ user friendly • Can't use if broken down • Dependent on handpump spare parts and maintenance tools/skills • Communal management
Rainwater harvesting (may be just to supplement others) **Posters W14,W15**	• Rainfall more than 3 months/yr, • Not economic where acceptable water quality within 40m of the surface	• Only alternative if groundwater saline or too deep for handpump • Can use roof area of schools • At low cost can reduce burden of water collection trips for part of the year • Rainy season is when women have least time to spare and people are most vulnerable	• Expensive per household if providing year round storage • Not good quality for drinking

FIG 4.1 POSTERS: RURAL WATER SUPPLY

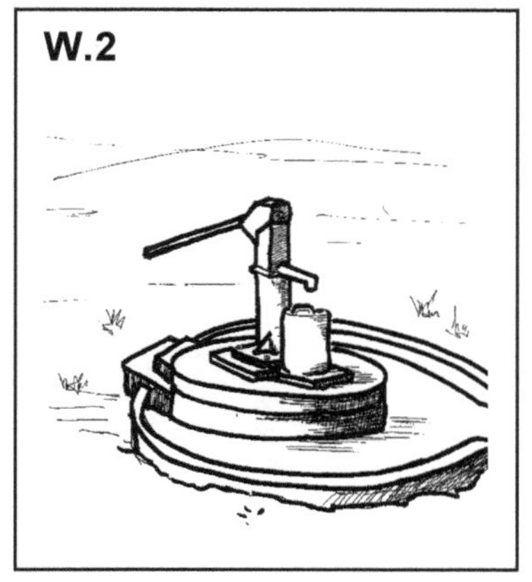

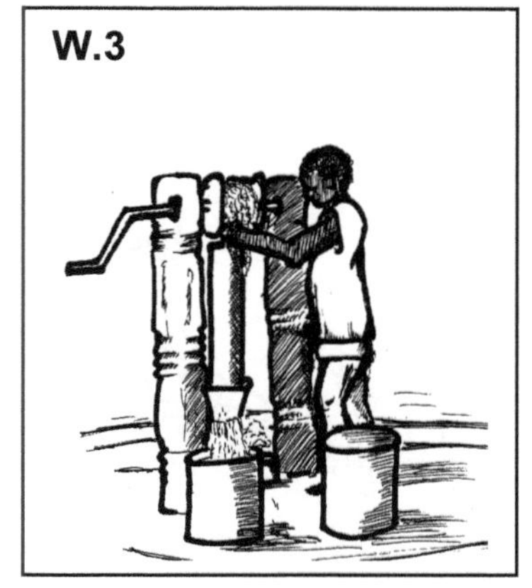

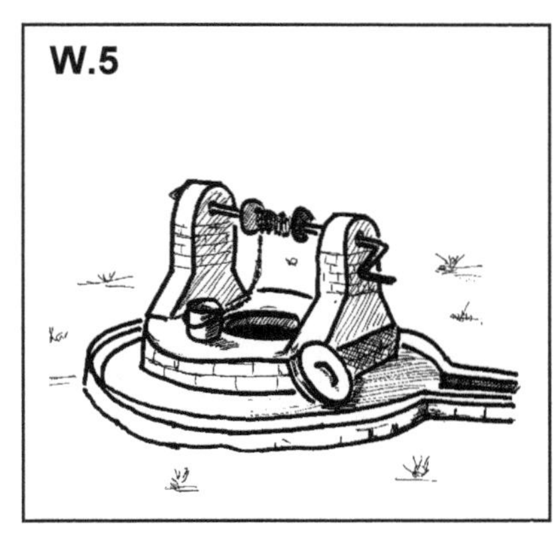

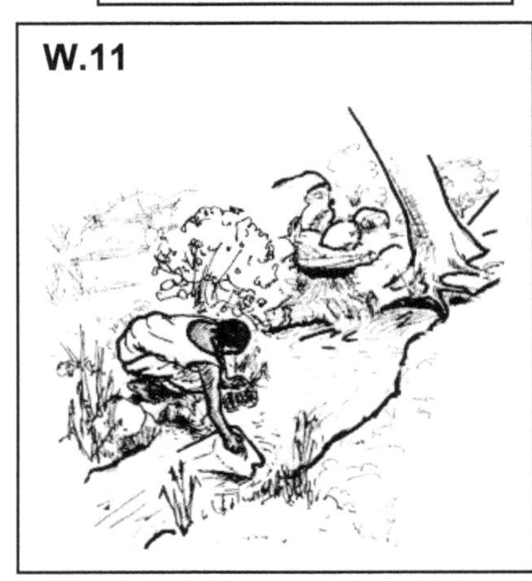

W.13

W.14

W.15

W.19

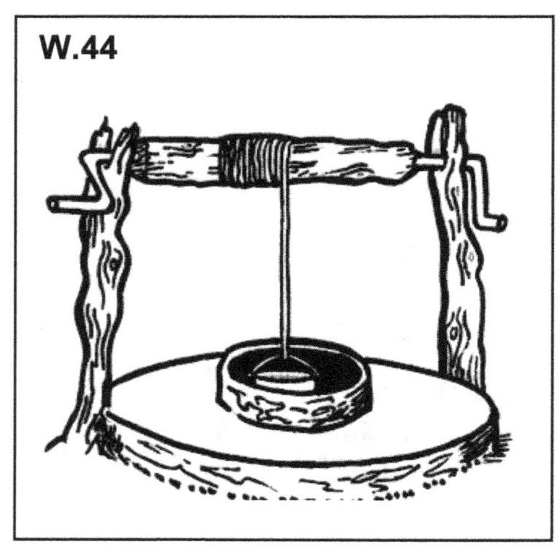

W.44

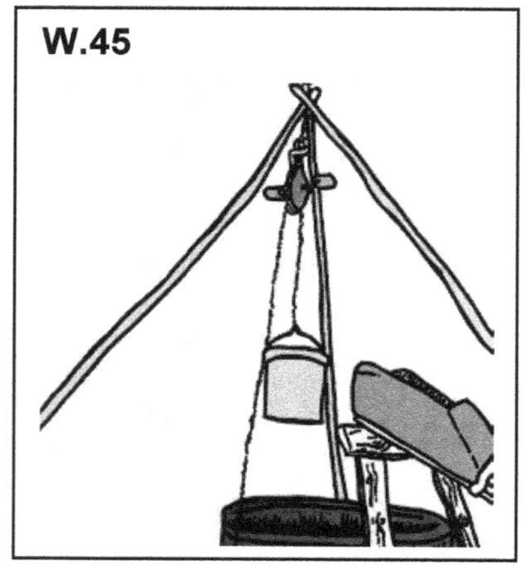

W.45

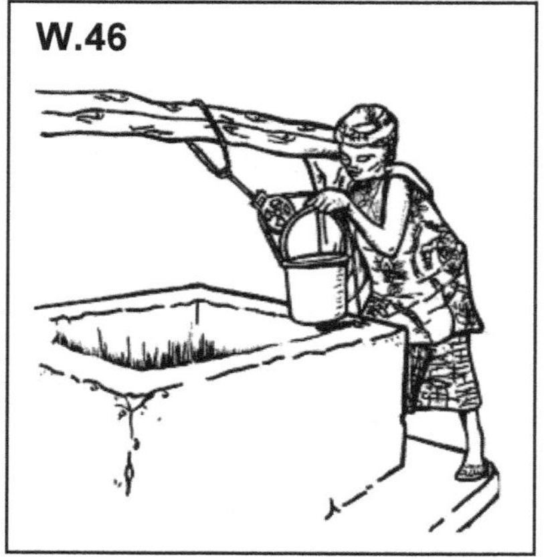

W.46

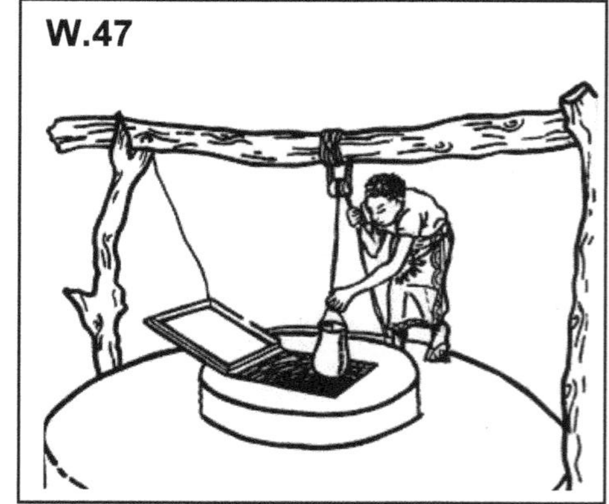

W.47

4.2.1 IMPROVING TRADITIONAL SOURCES (BACKGROUND INFORMATION FOR FACILITATORS)- see also manual - *Low Cost Water Source Improvements*, S.Sutton. (For details, see p.126 of this manual.)

Fig 4.3 Improvements to a family well and the effects:

BEFORE IMPROVEMENT, KAWANDA

AFTER LOW COST IMPROVEMENT, KAWANDA (INTERMEDIATE STAGE)

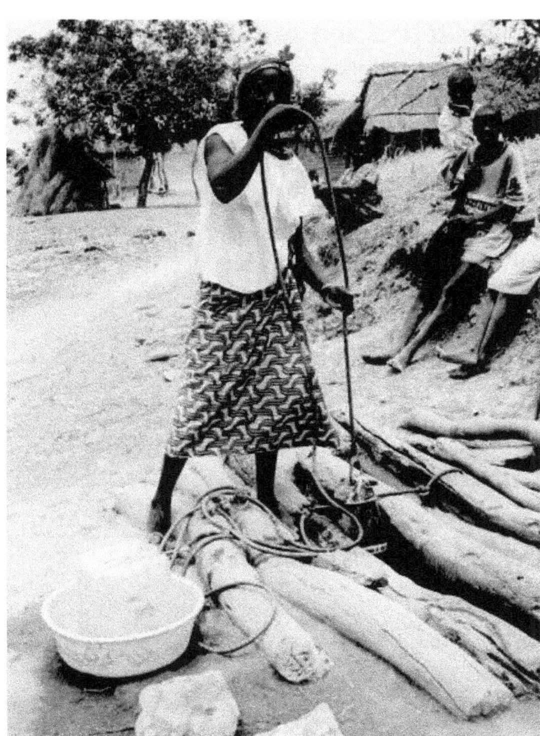

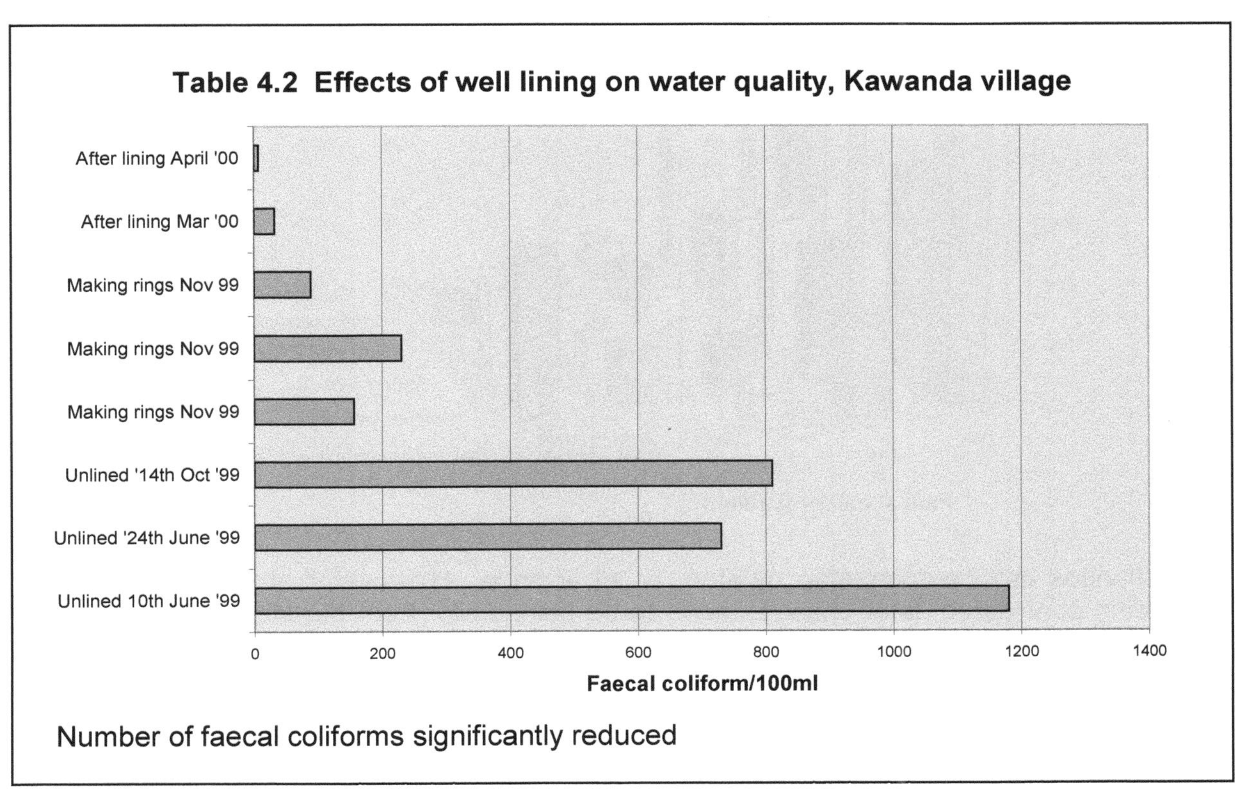

Table 4.2 Effects of well lining on water quality, Kawanda village

Number of faecal coliforms significantly reduced

Background information (continued)

INTRODUCTION   Most traditional sources (unlined wells, scoopholes, springs) can be improved significantly with small changes. For the   cost of two pockets of cement ($15) water quality can be improved significantly, and in most cases of high contamination, faecal coliform counts can be reduced a hundred-fold. These sources are still the main source of water for over 50% of the rural population.

The greatest improvement is made by lining the top of the well and making a small apron, with drainage. This also makes access easier to draw water, and can mean that a windlass or even a handpump can be installed, but access for deepening can be kept. This work (done by the users) can take just two or three days, but the results are very noticeable, often leading to more people coming to use the source. It also encourages others to make their own wells, as they see how nicely they can be completed and how useful they can be.   Family wells reduce distance to carry water, and allowing irrigation of small areas of high value crops. Amounts of water available in the home are increased, and the burden of water collection reduced. Similar changes can be made to reduce contamination and improve reliability and access to scoopholes and springs.

Table 4.3 Water quality in different source types

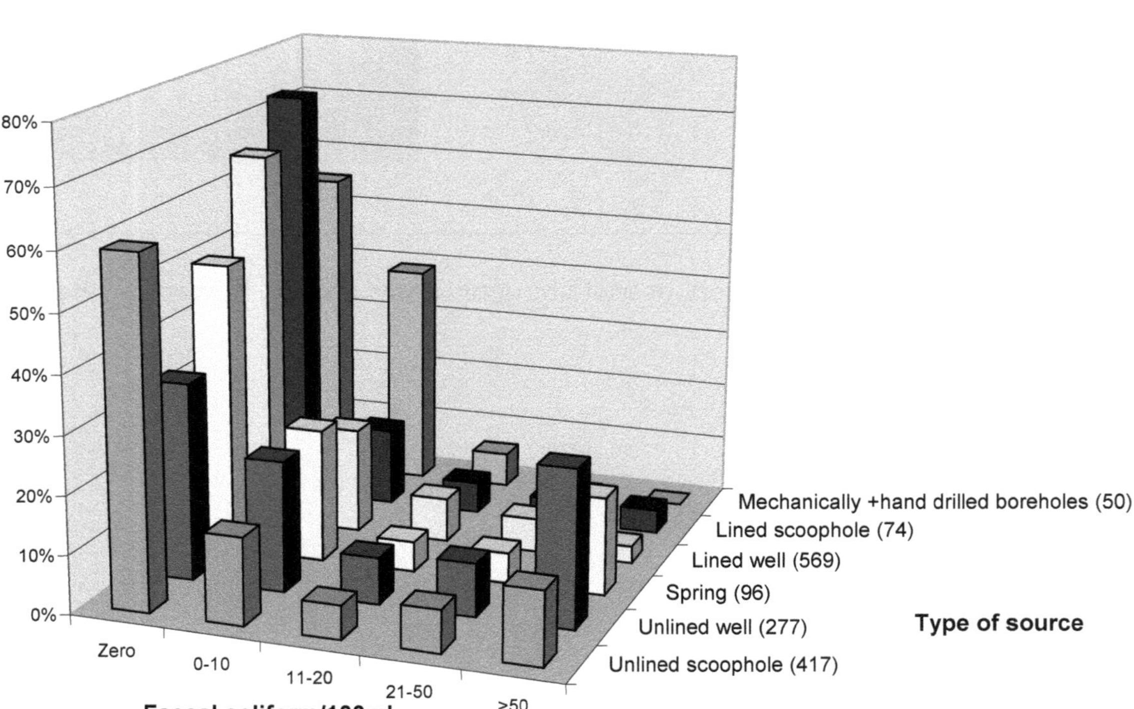

Changes can be done step by step, or all at once. However if staged up-grading is being considered, later changes need to be planned for from the start. For instance if there is a chance that, on shallow wells, a windlass or pulley may be wanted later, the poles on which it would be mounted should be installed before the apron is made, so that it does not have to be broken again. Similarly, the top slab should have an opening wide enough for a bucket and chain, so that the slab does not later have to be replaced.

4.2   ALTERNATIVE IMPROVEMENTS TO THE EXISTING SUPPLY

INTRODUCTION

On many occasions, especially among smaller communities (less than 20 households), the only option in the short to medium term (for the next five to ten years at least) may be improving and copying existing sources. This can be done i n ways that make the most use of local skills and materials. They can also be improved in stages so that finally the well may look little different from a conventionally protected well design, but should be better cared for, as the skills and resources to maintain it are local and t he same as those used in its construction. This session can follow on 2.1 "What is wrong with our water supply?" directly or preferably after sessions identifying resources, skills and community strengths.

PURPOSE   For the community to decide what   they would like their water supply to be like within three years and w hether they can complete improvements as one step, or whether they need to do it in stages.

MATERIALS

Posters of traditional source features/improvements, A4 drawing paper, pens/pencils, erasers.

TIME. 2 hours.

METHOD.

1. Take the drawing completed in 2.1 and ask members of the community to summarise the problems they identified. Ask also for the aspects of the source which people liked.

2. How can the problems be s olved? As far as possible get people to propose their own solutions. Then get them to sort the pile of drawings which feature possible solutions into those they regard as preferred, preferred but too difficult, and not   relevant or not preferred.

3. Discuss the preferred changes first and ask them to work out what would need to be done. Fol low this with those which are preferred but too difficult. Discuss the reasons why they are suitable against the reasons why they are too difficult. Is everyone agreed, or do some feel that greater changes could be made if the work were split into stages or outside help was called for.

4. In the second part of the discussion, divide the participants into a m aximum of three groups and ask each to draw how they think the source should look after three years, compared to now. Note the features and changes they want to make.

5. When they have finished, ask them to explain, and s ee whether, if there are differences in drawings, they can agree on a 'final design'. Ask that the person who has kept the 'existing situation' drawing safe, should do t he same with the new one. Summarise the changes that have been proposed. During the discussion, offer technical solutions from your own experience if necessary.

The value of demonstration / pilot wells

If others see a nice well which is kept cleanly and is popular to use, they can think whether they would like to try and make the same. In Northern Province in Zambia, it was found that when one person upgrades their well, neighbours are encouraged to dig their own, as they see that, over time, they can end up w ith a s upply which belongs to them, an investment from which they directly benefit, and a facility of which they can be proud.

DISCUSSION POINTS

- If there are several sources, how many will be improved?

- Will sources need deepening/ cleaning out before other headwork improvements are made?

- What parts need c hanging? Consider partial or full lining, top slab, lid/ cover, apron, lifting device/ pole to keep rope clean, drainage, soakaway, and diversion of run-off, plus possibility of use for irrigation.

- What are the benefits the proposed changes are predicted to bring, and who will benefit?

- What is the ownership of the source(s). Are owners and users both happy with the changes and prepared to contribute whatever is necessary, including their labour?

NOTE TO FACILITATOR:

- Especially where family wells are concerned, rather than emphasising only the health benefits, discuss also the convenience and f inancial benefits which can come from changes.

- Refer also to background information in 4.2.3. This shows the main problems users often identify with traditional sources, and the improvements which can help solve them.

Relevant Posters	
Lining shafts	W29-36
Top of the well	W2, W4-7,9,12 and W39-47
Lifting devices	W1-7, W16-18, W44 -47
Drainage	W7, 41, 42

Fig 4.2 RESULTS FROM DISCUSSION: NOW AND NEXT YEAR

EXISTING SITUATION

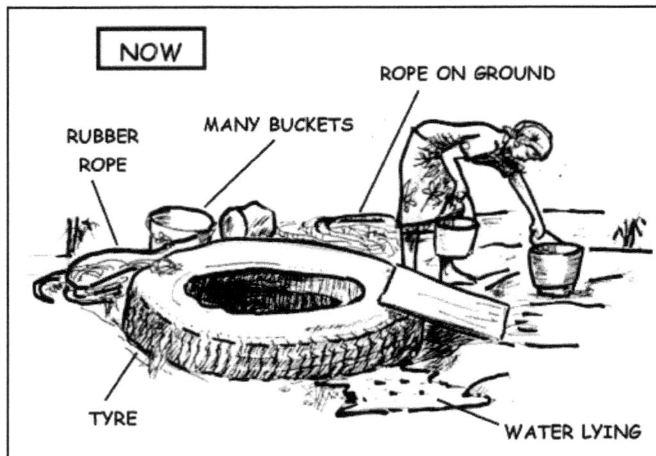

Problems. Water-logging, bucket gets muddy, makes water dirty, dangerous for children when lid not closed, ropes get dirty, dirt gets into water making it cloudy. Fear that water around may make top unstable and cave in.

PLANNED

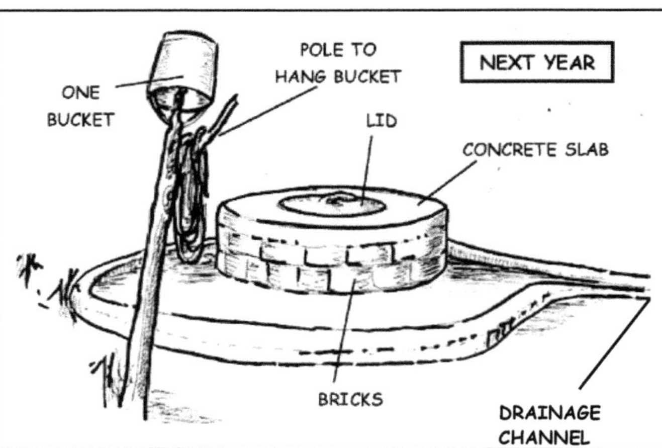

Solutions. Build up surroundings, set bricks from 0.5m below ground level, make brick apron and drainage channel, make top slab with small opening and well-fitting lid, set pole to hang bucket, one bucket only. Make and keep to rules to make surroundings clean and safe.

4.2.1 IMPROVING TRADITIONAL SOURCES (BACKGROUND INFORMATION FOR FACILITATORS)- see also manual - *Low Cost Water Source Improvements*, S.Sutton. (For details, see p.126 of this manual.)

Fig 4.3 Improvements to a family well and the effects:

BEFORE IMPROVEMENT, KAWANDA

AFTER LOW COST IMPROVEMENT, KAWANDA (INTERMEDIATE STAGE)

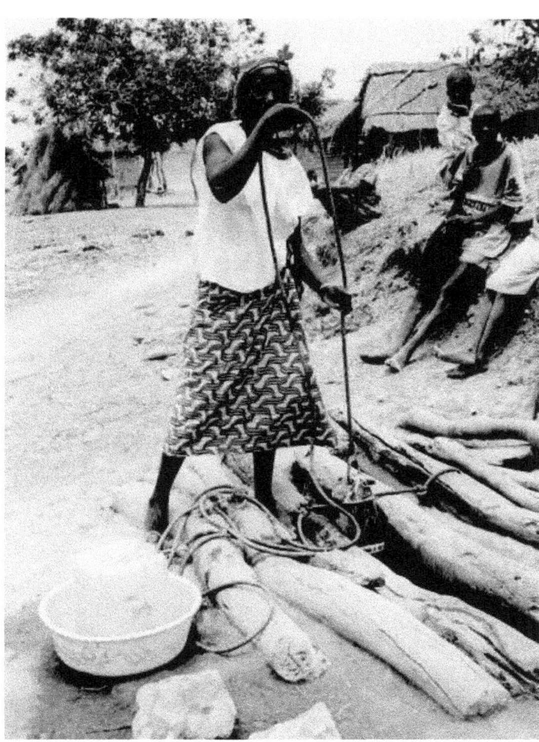

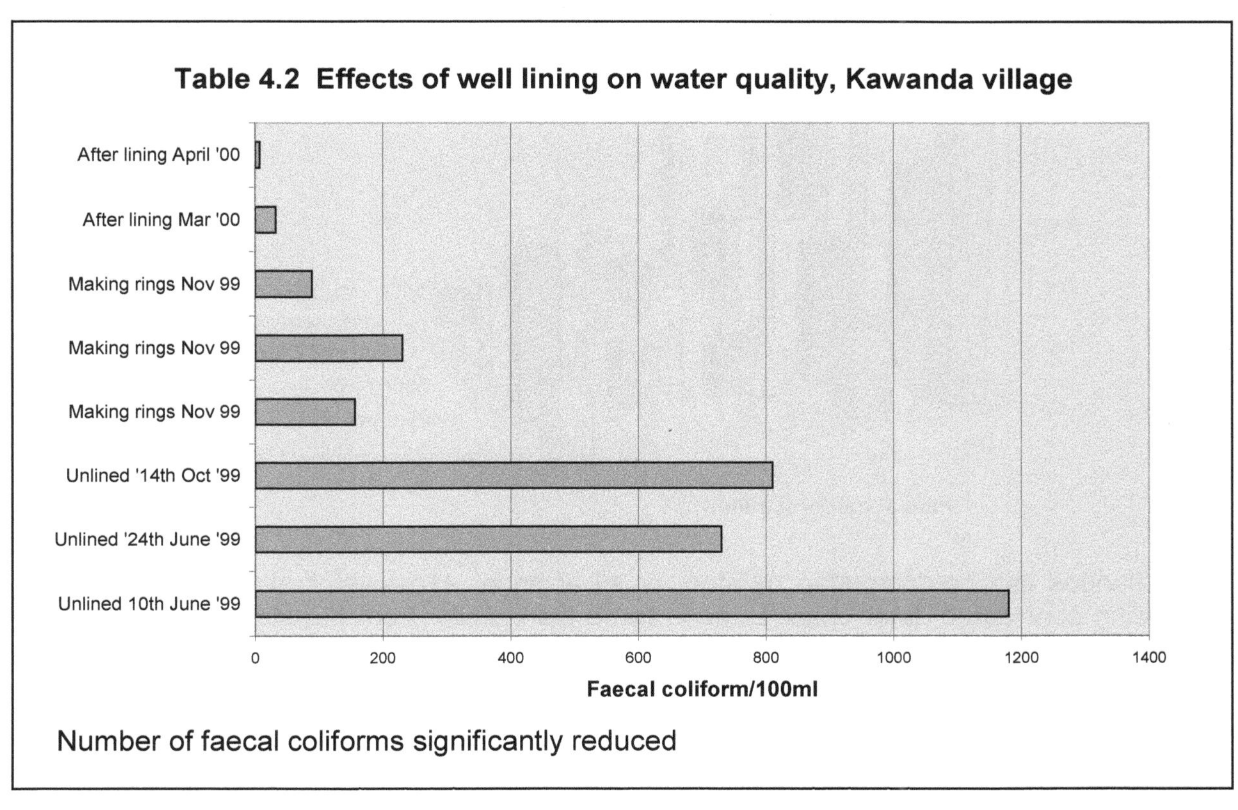

Number of faecal coliforms significantly reduced

Background information (continued)

INTRODUCTION   Most traditional sources (unlined wells, scoopholes, springs) can be improved significantly with small changes. For the   cost of two pockets of cement ($15) water quality can be improved significantly, and in most cases of high contamination, faecal coliform counts can be reduced a hundred-fold. These sources are still the main source of water for over 50% of the rural population.

The greatest improvement is made by lining the top of the well and making a small apron, with drainage. This also makes access easier to draw water, and can mean that a windlass or even a handpump can be installed, but access for deepening can be kept. This work (done by the users) can take just two or three days, but the results are very noticeable, often leading to more people coming to use the source. It also encourages others to make their own wells, as they see how nicely they can be completed and how useful they can be.   Family wells reduce distance to carry water, and allowing irrigation of small areas of high value crops. Amounts of water available in the home are increased, and the burden of water collection reduced. Similar changes can be made to reduce contamination and improve reliability and access to scoopholes and springs.

Table 4.3 Water quality in different source types

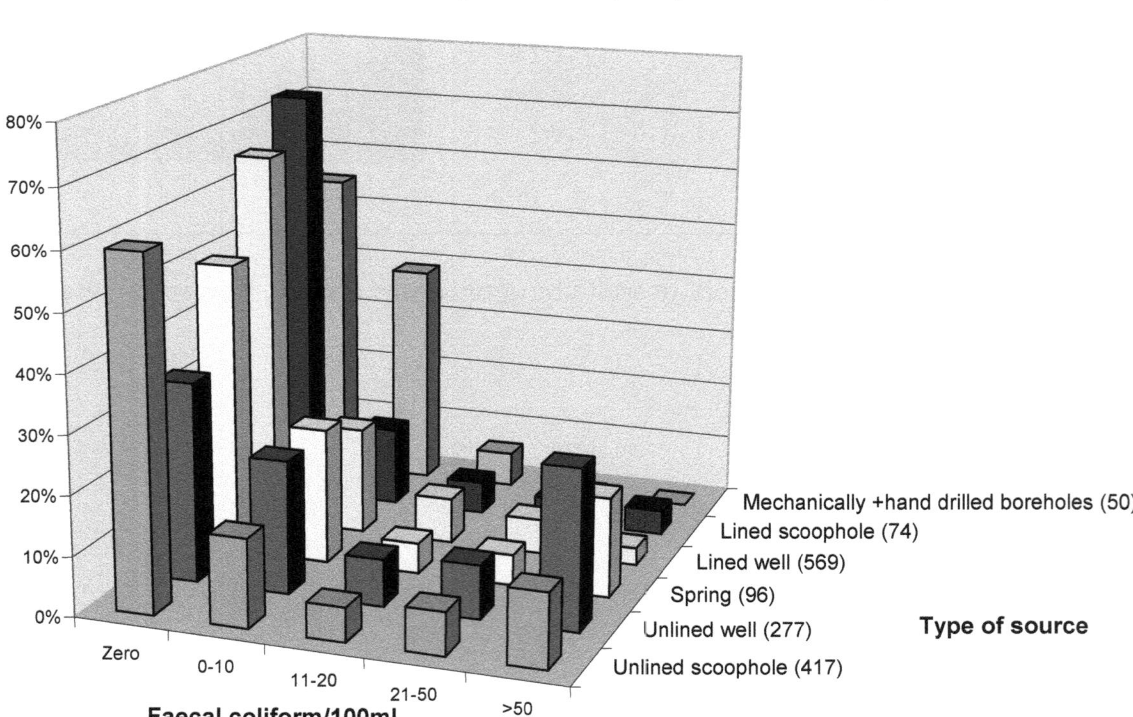

Changes can be done step by step, or all at once. However if staged up-grading is being considered, later changes need to be planned for from the start. For instance if there is a chance that, on shallow wells, a windlass or pulley may be wanted later, the poles on which it would be mounted should be installed before the apron is made, so that it does not have to be broken again. Similarly, the top slab should have an opening wide enough for a bucket and chain, so that the slab does not later have to be replaced.

> **Step by step improvement (could be over several years) for unlined well, progresses as economic situation improves.**
>
> 1. Lining top 1.5 metres, whole shaft, or top and bottom only
> 2. Adding top slab, lid and poles
> 3. Making an apron and drainage channel / soak-away
> 4. Windlass or pulley
> 5. Construction of cistern for irrigation.
> 6. Replacement of windlass by low cost handpump
> 7. Replacement of handpump by windmill or solar pump
> 8. Piped supply to house

Surveys in Zambia have shown that rivers and ponds tend to be the most contaminated sources followed by unlined wells. Lined wells and boreholes generally offer the best quality.

In all cases, as with standard lined wells with bucket and windlass, if users do not have clean hands, water may be contaminated. It is important to combine any improvements to sources with discussions on hand-washing, and whether a hand-washing device at the source would have any benefit. Problems often arise because people do not draw water until there is none left in the house, and cannot therefore wash their hands before handling collecting vessels. The situation is made worse where people put the carrying container directly into the source, rather than using a special bucket or scoop. The greater the turnover of water, (ie., the more water that is used,) the more likely that the quality will be good. Whilst handpumps can provide higher quality water, this needs to be weighed against their lack of reliability and cost of maintenance.

NB. Source improvements should only be considered on wells which are more than two years' old, (preferably older), and which have not gone seasonally dry. It may be more difficult to deepen improved wells, and almost impossible if they are fully brick-lined.

Well chlorination: Encourage users to do this if water goes cloudy or tastes bad.

> **STEP 1** Explain to users what is going to be done and why. Next time they will carry out the procedure themselves
>
> **STEP 2** Measure the depth of water in the well, using the bucket and rope. Bail out until there is less than an arm's length of water in the well
>
> **STEP 3** Keep three buckets of water by the well
>
> **STEP 4** Pour one 750 ml bottle of Jik or other laundry bleach into a bucket almost full of water. Pour this bucket into the well, followed by the other two buckets of water to ensure good mixing.
>
> **STEP 5** Leave the water for a minimum of one hour (preferably two) and someone at the well to warn users not to take water.
>
> **STEP 6** Bail water out again, using the bailings to wash around the well. Continue until the smell of chlorine has gone, or water taste is acceptable to users.

Table 4.4 BACKGROUND INFORMATION ON IMPROVING TRADITIONAL SOURCES

UNLINED (FAMILY) WELLS	ADVANTAGES	
Advantages of increasing their numbers	• Convenience- shorter distance for carrying bulk water for washing etc • 'Private' supply, real benefit from own investment • Possibility for income generation from seedlings/veg crops • Improved environment, cool attractive surroundings to house • The first step of several to make a convenient, safe and usually reliable supply • Allows an individual to have control over his investment, and over management decisions	
Advantages of improving them	• Improvement in quality so that they can also provide safe drinking water, reducing distance for collection • Greater safety, reducing risk of collapse or children falling in. • Possibility of more permanent headworks, including pulley or windlass or low cost pump, and use for irrigation • May also improve reliability over time	
Improvements which have most impact on quality (Q) and reliability (R)	1. Lining (concrete or brick) partial or full Q+R 2. Reduction in the water ponded around (having an apron and drain) Q 3. Ability to abstract more water (windlass or pump) Q (R?) 4. Building up parapet above ground level.	
SCOOPHOLES & SPRINGS	**ADVANTAGES**	
Advantages in increasing numbers	• Reduce queuing • May reduce distance • Allow greater control over use and management	
Advantages in improving them	• Improves reliability and ease of access • Provides cooler water, generally of better quality • Stops insects/ small animals from dropping in • Strengthens management • Reduces need to dig and move to sources which are further away in the dry season	
Most effective improvements	• Lining (Q+R) • Cover (Q) • Drain (Q) • Bailing	• Hand washing device • Place to put containers out of dirt

NB. Single scoop has little effect. It seems that the care people give to keep their own scoop/ bucket clean may counterbalance the expected advantages of having one communal one.

Table 4.5 MAIN PROBLEMS AND SOLUTIONS - TRADITIONAL SOURCES

FREQUENTLY VOICED PROBLEMS	MINIMUM PREVENTION	BEST OPTIONS
1. Rainwater/ run-off brings dirt into the source	1A.Making an earth bank around the source, and/ or diversion channel for run-off and spillage *(See Posters W40, W41)*	1B. Raise opening of source above ground level, by lining and providing apron. *(See Posters W41, W42)*
2. Animals drink from it/ snakes/ leaves fall in	2A. Provide cover and raised edge such as old tyre or wheel hub *(See Fig. 4.2)*	2B. As above, with top slab and cover *(See Posters W40, W41)*
3. Sides collapse	3A. Line section which collapses *(See Posters W32, W33)*	3B. Line whole depth *(See Posters W30, W31, W34)*
4. Access is muddy, containers get dirty,	4A Provide step or apron on which containers can be set *(See Posters W4, W5, W7)*	4B. Provide drainage channel and soak away *(See Posters W41, W42)*
5. Water is dirty/ cloudy	5A Provide all the above, bail out water when necessary	5B. All the above, increase turnover of water by bailing or increased water use (more users or irrigating crops.)
6. Source dries up sometimes	6A Deepen as long as walls will remain stable	6B. Deepen with concrete rings to support sides. (Not usually necessary to line whole of deep well.)
7. Surroundings are muddy and rules not kept to	7A 1-4 above, plus strengthen/ form management structure and encourage community to set rules	7B. 1-4 above, plus strengthen/ form management structure and encourage community to set rules by initiating discussions on environmental hygiene/ sanitation.

Fig 4.4 BASIC ALTERNATIVES FOR LINING SHAFTS (FOR WELLS, SCOOPHOLES & LATRINES)

W.29 Shallow brick-lined, for spring/scoophole or latrine pit

W. 30 Fully brick-lined well with concrete rings below highest water level

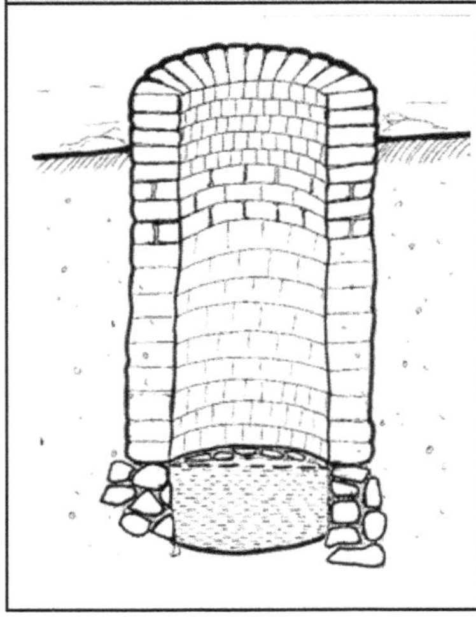

W. 31 Fully lined well, concrete block or brick and stone, with 'dry stone' below water

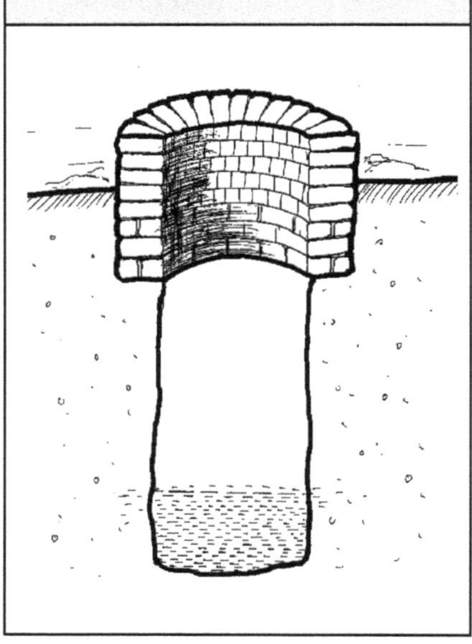

W.32 Partial brick lining, old or new wells

VIEW FROM THE TOP

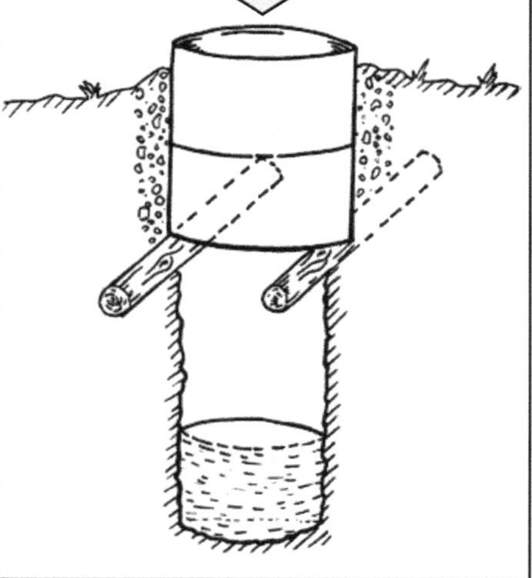

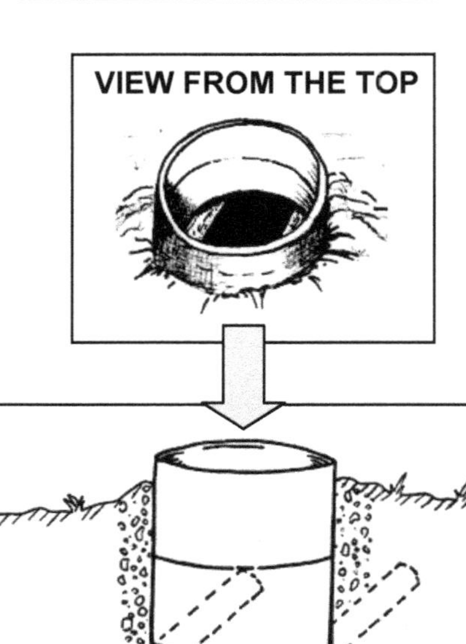

W.34 Full lining with concrete rings

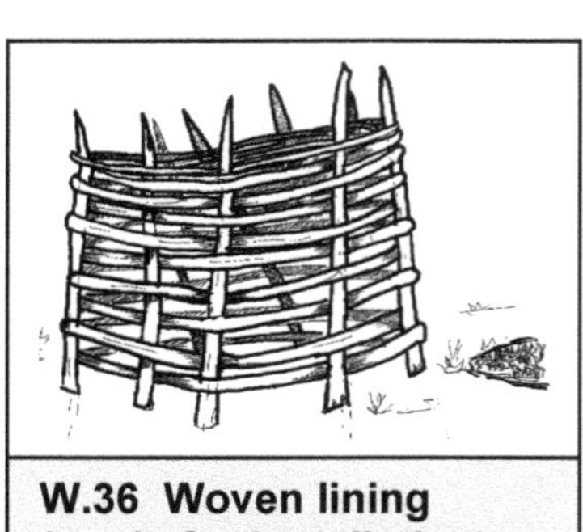

W.35 Partial wood lining for scoophole, well or latrine

W. 33 Partial lining with concrete rings (in old large diameter well)

W.36 Woven lining (ready for installation)

74

Fig 4.5 MINIMUM OPTIONS FOR THE TOP OF SOURCES, STEP BY STEP IMPROVEMENT (Try to get participants to make their own drawings relevant to the source being discussed.)

W.37 Unimproved scoophole/ well

W.38 Mouth of shaft built up (mound) to keep out run-off

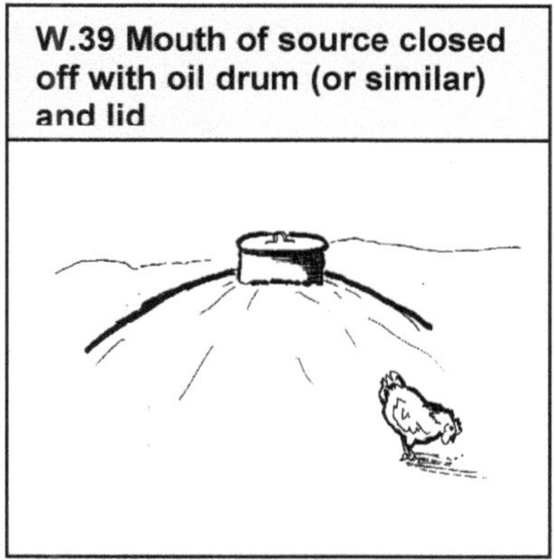

W.39 Mouth of source closed off with oil drum (or similar) and lid

W.40 Top strengthened with brick parapet, and small apron to remove spilt water

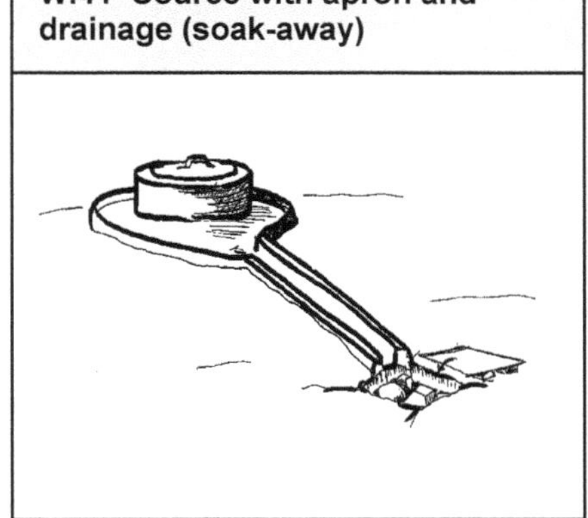

W.41 Source with apron and drainage (soak-away)

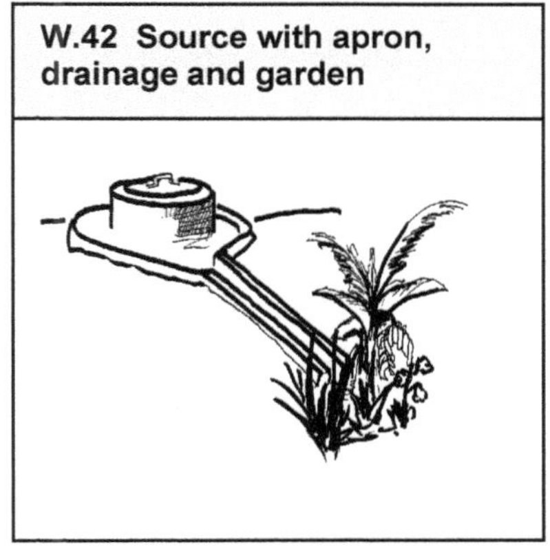

W.42 Source with apron, drainage and garden

THEN PROGRESS using options such as those below or W.2, 4, 5, or 6

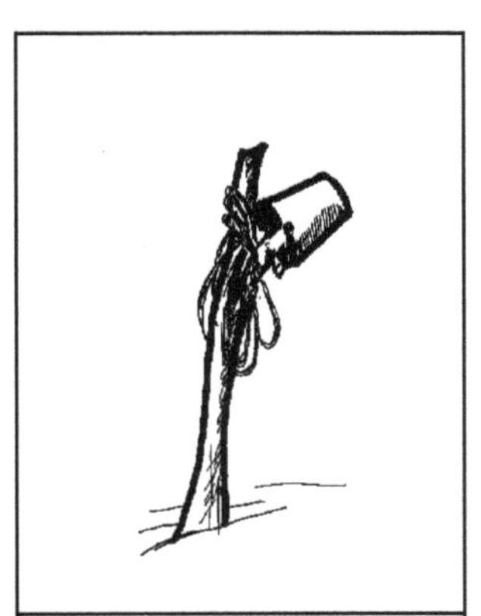

W. 43 Pole to hang bucket and rope

W.18 Treadle pump

W.17 Low cost (rope) hand-pump

W.1 Standard handpump, usually Afridev or India Mk II.

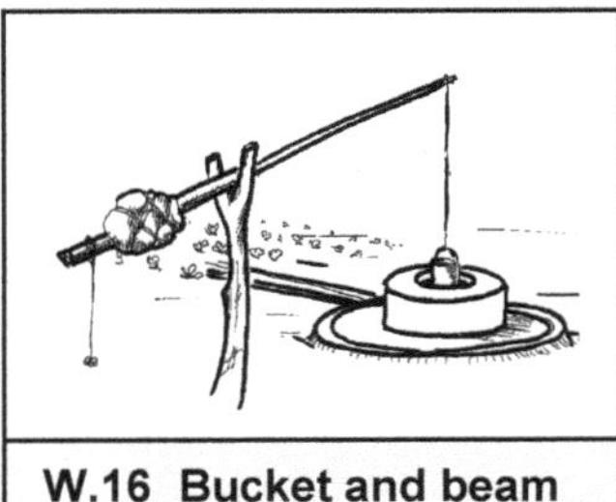

W.16 Bucket and beam

W.44 Locally made windlass

W.46 Pulley

4.3 TECHNOLOGY OPTIONS FOR IMPROVED SANITATION

INTRODUCTION In Section 2.2 the community identified a number of advantages and problems associated with their current defecation and s anitation practices. These issues will need to be addressed as part of the community's overall plans to improve the water, sanitation and health status of their village. Some of the solutions to the problems identified may relate to changing behaviour but it is likely that some relate to the construction or improvement of latrines. Communities and individual households will need to consider many different aspects when deciding upon the changes they want to make to the sanitation situation in their village. They will need to draw upon their experience both of existing facilities in the village and those that they have come across elsewhere or have heard about. The facilitator is likely to have to be able to answer many questions from the community and for this purpose background information about sanitation options is attached. The session utilises the Sanitation Ladder with Pair-Wise Ranking and Focus Group Discussion tools.

PURPOSE

This session aims to provide communities and hous eholds with enough information to allow them to make well informed decisions on t he sanitation technology they wish to adopt to address the problems they have previously identified.

TIME 2-3 hours

MATERIALS

Posters showing different sanitation options; plastic folders; counters / stones; paper and marker/drawing pens.

TARGET GROUP(S)

Community group as a whole; smaller mixed groups; individual households.

METHOD

1. Explain the purpose of the session to the whole group.

2. Summarise the conclusions of Session 2.2 which identified the advantages and problems of the community's current defecation practices and any relevant outputs from sessions in Section 3 (Preparing the Community to Solve Problems).

3. Explain briefly what type of sanitation each poster of the Sanitation Ladder shows, but **not** the advantages and disadvantages of each.

4. Ask participants to pick out the ones they are familiar with and what they know of each, both problems and good points. Additional types of latrine known by the participants but not included in the posters can be dr awn by the group and add ed in to the ladder. Unfamiliar models will h ave to be explained to the participants if they are regarded as relevant. Care should be taken not to 'push' any one system but to respond to questions honestly in relation to how each functions, what materials it requires, how often they are found in the area, and the relative cost of each.

5. Divide participants into smaller groups and ask them to consider which they regard as the worst option and which they consider to be the best option. Leave groups to make their own definitions of 'best' and 'worst'. Each group presents its findings and then the whole group is asked to combine the different ladders into one ladder.

6. The facilitator should answer queries and note down points of interest (advantages, disadvantages) that are brought up dur ing the discussion as participants consolidate their conclusions into one ladder.

7. Ask participants to go home and discuss what they have learnt with others in the family. At the next (short) meeting they will be as ked to outline what they feel they could change as

individual households, and w hat they would like to see changed as a community.

DISCUSSION POINTS

- The community's perceptions of how each latrine type works

- The community's perception of the strengths of each type of sanitation

- The community's perception of the weaknesses of each type. What are the constraints to people building latrines of different types.

- What the community thinks the cost of each latrine type is and how much of this they are likely to have to pay.

- How well the different models fit in with existing customs and habits.

- The community's definitions of 'best' and ' worst', and which alternatives each would prefer.

NOTE TO FACILITATOR:

- Many people are reluctant to discuss what they do a bout going to the toilet and are embarrassed to talk about such things in public. Therefore this session should be run sensitively and with respect for people's feelings on this issue.

- The facilitator must take care during the session to avoid promoting only one (their preferred) technology option at the expense of all others. Offering communities a r ange of choice in types of latrines is **extremely important** as 'one size does not fit all'.

- The next meeting (Module 5.3) will require as many household representatives to be present as possible. Ask the headman to encourage people to attend, and ask participants to discuss the conclusions of this session (4.3) with those who did not attend.

- Before the next session the facilitator will need to check whether:

  - Funds for various technology options might be available from any local water and s anitation committee.

  - there are any NGO (or other agency) activities which might support sanitation development within the community.

EXAMPLE

The people of Choobana village in Monze district, Zambia, identified a lack of toilets in their village as a key reason why there were high numbers of diarrhoea and dy sentery cases. A decision was made by the community to change this situation and all households without a t oilet agreed to build a ne w latrine. The c ommunity was not interested in small (0.6m x 0.6m) Sanplat latrines as there had been experiences in the villages of these slabs breaking after construction. The extension officer for the catchment therefore discussed a range of 6 other latrine options with the community: a traditional pit latrine, an ordinary pit latrine, a l ow cost VIP latrine and t hree different types of ecological sanitation latrines. The advantages and di sadvantages and the material and l abour requirements of each type of latrine were discussed. At the end o f the session, all households had s elected an opt ion that they felt most suited their requirements. The majority of these were traditional pit latrines or ordinary pit latrines, but a few households chose to build VIP latrines and t wo were interested in trying out one of the ecological sanitation options. T he community agreed to meet with the extension officer again to work out what steps would be involved in building their new toilets and w hen these would happen.

NEXT STEPS	
If changes in handwashing are also associated by participants with changes in sanitation	Go to Module 4.7
If changes in personal hygiene are raised as associated issues by participants	Go to Module 4.9
If household hygiene is also to be improved	Go to Module 4.8
If no other changes are to be made	Go to Module 5.3

Suggested posters S1-S22

RECORD KEEPING

Note communities' views of different technologies, preferred options, and reasons.

Table 4.6 SANITATION TECHNOLOGY OPTIONS (BACKGROUND INFORMATION) Some ideas, but the community's views on advantages / disadvantages are more important. Their beliefs will determine whether they build and use a new latrine or not

TYPE OF SANITATION	ADVANTAGES	DISADVANTAGES
Open Defecation (Going in the bush, stream/lake) **Materials needed:** *None* *Posters S.11-S.15*	• Free materials • Available almost everywhere • Well understood • Common practice	• Unpleasant to see and smell • Helps fly breeding • Can lead to worm infections and transmission of other diseases • Liable to pollute surface water sources • Lack of privacy • Distance from house, especially during sickness /at night
Cat Method (Going in the bush as above, but burying faeces) **Materials needed:** *Hoe* *Posters S.16 and S.17*	• Hoe available in most households • Possible almost everywhere, even rocky areas • Common practice especially with children's faeces • No maintenance costs • Lower risk of worm infections	• Faeces near ground surface, may be disturbed by animals and washed away by heavy run-off, or carried by feet • Smell reduced but not removed • Lack of privacy • Can lead to worm infections • Faecal matter can be carried away on feet.
Traditional Pit Latrine (Walls of mud/ grass/sacking, no roof, unlined pit with timber support at top) **Materials needed:** *Wooden poles, earth/clay, grass* *Poster S.1*	• Cheap to construct - uses local materials • Easy to build -similar to local houses • Well understood • Isolates excreta from the environment • Can be upgraded • Few Operations & Maintenance requirements • More private • Can be near house	• Often smells, flies can breed in pit • Not easy to keep 'slab' clean • May have a lid for the squat hole which can be a source of contamination as people replace lid before washing hands • Liable to rotting/termite attack • Construction materials not very long lasting unless treated • Unprotected from rain • Tendency for pits to collapse when unlined • Possible groundwater pollution if near water point • Difficult to construct in rocky areas or areas of high water table • Latrine abandoned when pit full
Improved Traditional Pit Latrine (Roofed, grass / mud / sacking walls, pit lined or unlined, squat hole lined with clay/plaster/mortar or 0.6m x 0.6m square slab with foot plates and well fitting lid to cover squat hole) **Materials needed:** *1/4-1/2 pocket cement; re-bar: hole cover; wooden poles, earth/clay, grass* *Posters S.2, S.4, S8*	• Low cost - uses few non-local materials • Builds on local knowledge/practice • Isolates excreta from the environment • Slab easier to clean and some can be re-used • Flies and smell are reduced • Can be upgraded • Few Operations & Maintenance requirements • Gives privacy	• Often smells, and flies breed in pit • Requires a lid for the squat hole which can be a source of contamination as people replace lid before washing hands • Liable to rotting/termite attack • Construction materials not very long lasting unless treated • Tendency for pits to collapse when unlined • Potential for groundwater pollution if near water point • Difficult to construct in rocky areas or areas of high water table • Latrine abandoned when pit full

TYPE OF SANITATION	ADVANTAGES	DISADVANTAGES
Ordinary Pit Latrine (Sanplat or larger top slab, walls of mud/ burnt brick, roof of grass or zinc) **Materials needed:** *1/2 -1 or 2 pockets cement; rebar; Superstructure materials; hole cover; roofing materials; pit lining material* ***Posters S.5/6 and S 20/22***	• Easy to understand/builds on local practice • Easy to build and is longer lasting • Isolates excreta from the environment • Concrete slab easy to clean and is re-useable • Can be upgraded • Can be used in schools, clinics etc. • Gives privacy	• More expensive • Often smells, and flies breed in pit • Requires a lid or cover for the squat hole, this can contaminate hands • Tendency for pits to collapse when unlined • Potential for groundwater pollution when pits are deep • Can be more complicated to build if un-reinforced (domed) • Difficult to construct in rocky areas or areas of high water table • Latrine abandoned when pit full
Ventilated Improved Pit Latrine (VIP) (Semi-permanent building, concrete slab, designed to reduce fly breeding and contamination, pit usually lined) **Materials needed:** *1 - 6 pockets cement; rebar; vent pipe; fly screen; bricks for Superstructure; pit lining materials; roofing materials* ***Posters S.3, S.7, S.9***	• Few smells or flies • Isolates excreta from the environment • No cover for squat hole needed • Concrete slab easy to clean and is re-useable • Possible to include twin-pits and thus make 'permanent' • Can be used in schools, clinics etc. • Gives privacy	• Medium to high cost (depending on design) • More difficult to understand operation • More difficult to build well • Moderate maintenance requirements • Some designs use significant non-local materials • Dark/semi-dark inside the toilet • Potential for groundwater pollution when pits are deep • Difficult to construct in rocky areas or areas of high water table • Latrine abandoned when pit full and investment lost
Ecological Sanitation (Latrine with shallow accessible pit(s). Faeces kept dry, not mixed with urine or water, only with soil, or ash) **Materials needed:** *1/4 - 1 pocket cement; rebar; bricks/pit lining materials; Superstructure materials; roofing materials* ***Poster S.10***	• Can be low cost (depending on design) • Can be used in high water table/rocky areas • Few smells or flies • Isolates excreta from the environment • Designed to change excreta into fertiliser • No water needed • Shallow pits limit potential for groundwater pollution • Cover for squat hole optional • Concrete slab easy to clean and is re-useable • Possible to include twin-pits and so make 'permanent' • Can be upgraded and gives privacy	• Not a familiar technology • More difficult to understand operation as maintenance requirements are very different from other latrine types • Some designs are more complicated to build • Some designs are expensive and use significant non-local materials • Higher operation and maintenance requirements • Needs acceptance by users of human excreta as fertiliser, and willingness/ capacity to clean out pits.
Pour Flush Latrine (Latrine with water filled air-lock isolating squat hole from pit, faeces and urine swept into pit by water poured in) **Materials: needed:** *1- 3 pockets cement; rebar; pit lining materials; super-structure and roofing materials*	• Smells and flies prevented by water seal • Isolates excreta from the environment • No cover for squat hole needed • Concrete slab easy to clean and is re-useable • Can be used in schools, clinics etc. • Gives privacy	• Medium to high cost (depending on design) • More difficult to understand operation • More complicated to build and maintain, especially water seal • Moderate operation and maintenance requirements • Uses significant amounts of water for flushing • Water seal can become blocked with anal cleansing materials • Once water seal is broken toilet becomes a wet Ordinary Pit latrine • Needs space for construction (which may be hard in urban areas) Pit may fill if high number of users, and problems of construction in rocky areas or high water table.

Fig 4.6 POSTERS: SANITATION - MAKE MORE OF YOUR OWN!

S.1 Traditional latrine (basic)

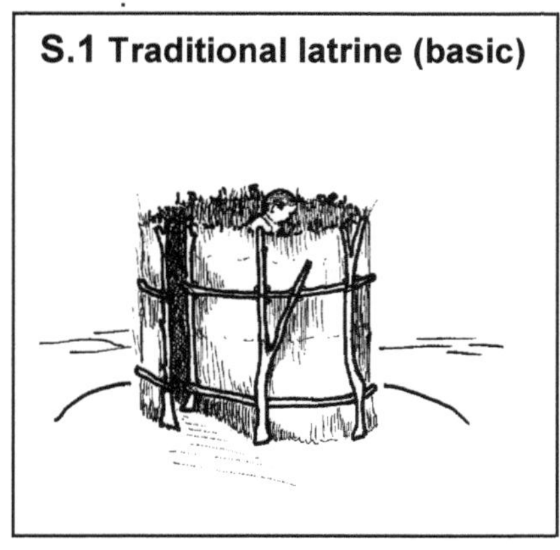

S.2 Traditional, improved (grass-roofed)

S.3 'Traditional' latrine, improved (VIP)

S.4 Traditional, improved

S.5 Ordinary pit latrine (1)

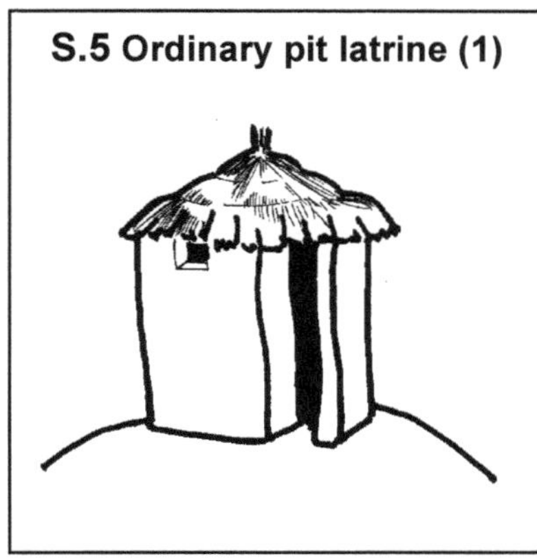

S.6 Improved pit latrine (2)

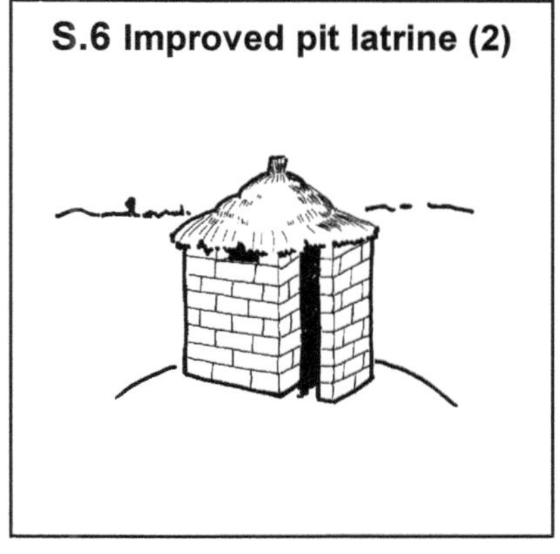

S.7 Brick/mud-brick improved pit latrine VIP

S.8 Improved pit latrine (3) with square SanPlat

S.9 Workings of a Ventilated, Improved Pit (VIP) latrine

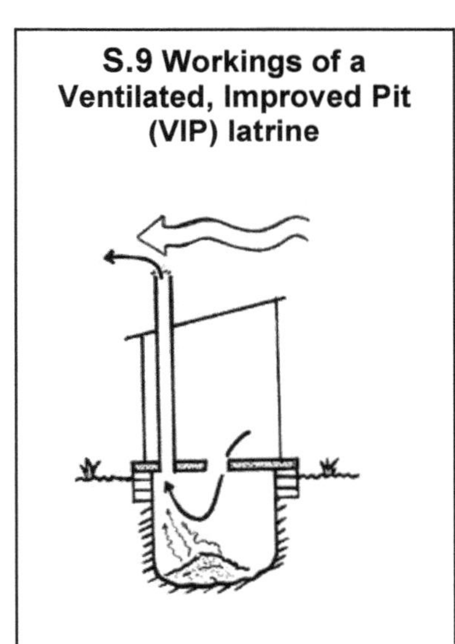

S.10 Ecological Sanitation/composting latrine

S.20 SanPlat

S.22 Wood slab for squat hole

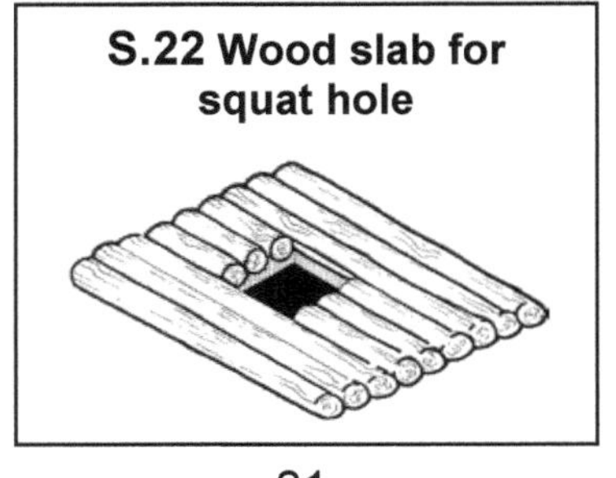

S.21 Clay / earth slab over wood

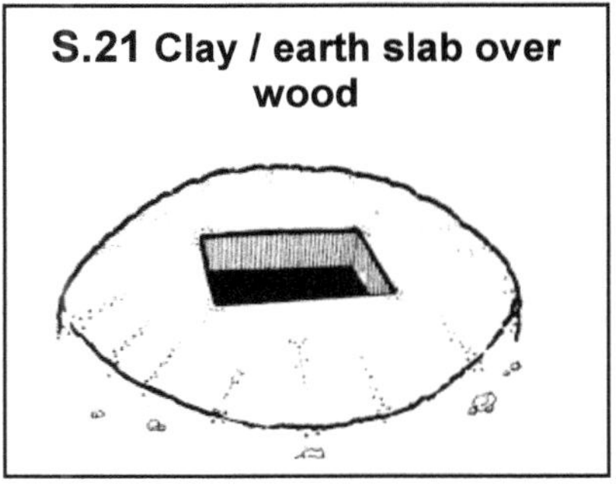

S.11 Going in the bush

S.12 Going in the lake, dam or river

S.13 Using a stream

S.14 Open defecation near houses

S.15 Going in the fields

S.16 Cat method 1

S.17 Cat method 2

4.4 DEVELOPING BEHAVIOURAL STANDARDS AT THE SOURCE

INTRODUCTION   Some of the problems identified in Session 2.1 relate to how people collected water and what other activities they undertook while at the source. They may also have related to how their children behaved while they collected water or waited for another member of their family to do so. This session uses **role play** and **posters of dai ly a ctivities** and can equally well be us ed for other protected source types.

Fig 4.7 Collecting water (example A): Jerrycan put into the water, standing in the water, run-off flows in, children around, good or bad practice?

PURPOSE   To establish what practices the community feel should be encouraged and what discouraged in order to maintain good quality water at the source.

TIME   2 hours

TARGET G ROUP   All those who collect water, particularly including young men and boys.

MATERIALS   Drawing of source problems from Module 2.1, posters of daily activities + defecation, pen, paper.

METHOD

1. If possible hold the meeting at the source (or a typical source if many).

2. Remind people of the relevant problems which came up i n Session 2.2 and what changes are proposed if Session 4.1 or 4.2 has been held.

3. Divide people into three groups, men women and children.

4. Ask each group to discuss what activities they undertake when coming to collect water, and how they draw water from the source. Ask them also to discuss what others (men and children for the women's group) do di fferently from themselves.

5. Ask the women's group to act what they do, perhaps with other members of their family accompanying them. (small children especially

6. Ask the other groups to do the same and the women to comment on how things differ.

7. Establish what they all feel are good practices and what are bad in activities around the source.

8. Discuss what other measures would help keep the source clean, and who will take responsibility to do them or see that they are done.

9. Using daily activity sheets, ask those who draw water to rank them in the order of those which most frequently lead to their having to go to draw water. (that is the activities which they are most frequently doing immediately before drawing water). Discuss which of these might most have a risk of faecal contamination to hands which may then draw water.

10. Get participants to agree on standards of behaviour / best practice for use of their water supply.

Suggested posters

G1-G14 Daily Activities / times

Plus some of **S11-17**

DISCUSSION POINTS

- What are the main ways dirt is getting into the water at the source, or in the container?

- How could it be m ade sure that hands and buc kets are clean before drawing or carrying water?

- How can children be en couraged to draw water in a s afe way? If they do not do s o now, what changes could be m ade to the well head to make it easier to draw water safely/ more difficult for them to dirty it?

RECORD KEEPING

Note down the changes that people propose and t he reasons for them. Propose another meeting to decide on r ules and who will ensure they are kept. If it is difficult to arrange meetings it may be pos sible to extend the present meeting to include this aspect.

Fig 4.8 Collecting water (example B): Water bailed into container, timber to stand o n, f lowing water, d ifferent practices, s ame si te as F ig 4. 7 b etter o r worse water quality?

FACTS FROM THE BASELINE SURVEYS

- Water stored in rural households in Zambia is very seldom found to be significantly more contaminated than the water source it came from. In most cases, keeping the source clean will mean clean water for the consumer.

- Most contamination of a water source with an opening above ground level, comes from dirty hands leaving dirt on the container which lifts out the water, or from the container being stood in dirt at home, in transit or at the source

---

NEXT STEPS

Go to Module 5.3 or 5.4 for action planning, or continue with Section 4 if household and personal hygiene have been voiced as problems.

EXAMPLES FROM THE FIELD

In Wayanda, Kaoma, none of the men showed that they washed the water container before filling it, but all the women did. The resulting water was thrown on the ground and formed dirty pools by the source, which users stepped in, bringing dirt. They all agreed that washing out the container was good practice, but no-one had previously thought to pour out the water further away or to make drainage, although all complained of the dirt.

At Kafula Chishika, Kasama, the community decided that no washing or bathing is allowed around the source, nor may rubbish be thrown or animals be allowed to roam nearby. Dipping of carrying containers in the source is forbidden and so is stepping in the water.

4.5 OPTIONS FOR INCREASING WATER AVAILABILITY AND USE IN THE HOUSE

INTRODUCTION In Session 2.5 participants looked at when there is a problem to have enough water in the house. This module suggests ways of promoting discussion on al ternative ways to increase availability and us e, using pocket chart voting and Focus Group Discussion.

PURPOSE To examine what changes at household level could improve water availability and use.

TIME 2 hours

MATERIALS Posters, pen, pencil and A4 papers.

METHOD

1. Summarise the common reasons found in Session 2.5 why there was sometimes not enough water in the house, or people felt they could not use water in the way they would like to.

2. Set out and explain a selection of posters which include ones of different ways to transport water, storage and washing containers, bathhouse, soap, hand-washing devices. Ask people to place counters (in the same way that people to select those which they think they did for Module 2.4) on the systems which they use at present and ask whether it is necessary to have water more easily available, and m ore effectively used.

3. Ask them if they can think of other things that would assist in this, and t o draw a poster representing each.

4. Take posters of different ways of collecting water and of increasing storage, and ask each group to discuss which are most relevant to them, and which could be done all groups.

5. Groups hold up t he posters of methods they use at present , and why they cannot use the others, giving their reasons. Others may comment on this, and then each group selects any additional posters that they think they could adopt in the near future. This should ideally include greater involvement of men in transporting water.

6. Summarise the conclusions drawn by participants on w ays of increasing use, and of making water transport easier and less of a bur den on women. Ask households to discuss this and s ee what behavioural changes they could make, and what materials/containers they might also need to purchase.

Suggested posters

W20-28.

Some of **H21-28**,

G3,8,9,11,12,13

DISCUSSION POINTS

- Water transport and i n what ways carrying water can be m ade easier. In what other ways can water be carried than on the head or hand, or by bicycle?

- Do most families have bicycles and do these need any modification to carry water?

Fig 4.9 Men/boys collecting water by b icycle i s an increasingly common sight.

- Has there been a noticeable change in who is seen collecting water at the source? Is it true for all families or have some so far not changed what they do? Do young men assist with water collection more than their fathers used?

- At what times of day could men help with water collection, to fit with their own activities?

- Would access to a bat hhouse make it easier for all members of the family to improve their personal hygiene?

- What changes do those households present at the meeting feel they could make to improve their situation?

<table><tr><td>NEXT STEP

Module 5.4 Making plans</td></tr></table>

NOTE TO FACILITATOR

If it does not come out in discussion, show, from the posters and pictures how much more men are now assisting women to collect water. It is no longer an activity which suggests weakness or lack of masculinity, but tends to be found in many progressive households and among younger men who welcome change and see the need for women to spend more time on c hild care and l ooking after the household.

Fig 4.10 Jerrycans are now plentiful in the market place

Fig 4.11 POSTERS: SOME WAYS OF TRANSPORTING WATER

4.6 ENCOURAGING GOOD HANDWASHING PRACTICE

INTRODUCTION In Sections 2.2 and 2.3 methods were described for looking at existing hand-washing practices. This used **pocket chart voting.** The results of those sessions can be used as a starting point for discussions on what changes people feel should be introduced.

PURPOSE To get participants to identify what is wrong with their present practices and in what ways they wish to change them. T hese changes should be des igned to block transmission routes identified in Section 1.4.

TIME 2 hours

MATERIALS Posters from Section 2.3 and not es from Sessions 2.2 and 2.3. Posters of hand-washing devices and/or materials to demonstrate construction of devices (see below)

METHOD

1. Review what was found in previous discussions on what people do at present.

2. Ask people to decide what they would like to be the most acceptable practice in terms of **WHEN** and **HOW** hands should be washed. This will require a repeat of the previous 'blind' voting, using the same posters, but asking for voting on how people feel they should act in future

3. On presenting the results, ask people to explain why they voted as they did. . What practices would they now regard as lazy or embarrassing and w hat is healthy, progressive, and makes one feel good?

4. Ask for any ideas of how water could be made more easily available for hand-washing by the latrine and in the house. O ffer additional ideas from below if necessary, of methods which allow good hand-washing without requiring large volumes of water.

5. Get commitments from each householder present and from the community leaders as to what changes they want to achieve within two months and within half a year.

Fig 4 .12 Handwasher: Just a g ourd a nd a stick are enough to make a handwasher.

DISCUSSION POINTS

- If people know that washing after latrine and before food are important, for what reasons do they not follow such a pr actice? How can this be overcome?

- Should people be embarrassed to be seen sharing one bowl of water for all the family? O r washing hands after going to the latrine.

- When should soap be used most?

- Should there be a facility to wash hands at the water source before drawing water?

- How can children be enc ouraged /persuaded to wash their hands more often and why is it especially important?

- Who are the most respected people and c an they be g ood examples for others to follow?

- What are the dangers of spreading dirt during handwashing?

NOTE TO FACILITATOR:

The aim is to encourage change in practices, but also involves changes in attitudes or beliefs. Demonstrate the changes in practice as far as possible so that others can see exactly what is meant. Show people examples of hand-washing devices and discuss with them that it may not mean carrying more water home, but just using the quantities available more efficiently.

Try to make sure you are seen doing the same yourself…. You are an important role model! Make hand-washing devices for your home and clinic today!

4.6.1   BACKGROUND I NFORMATION ON H ANDWASHING (FOR FACILITATORS)

Some ideas of handwashing devices to put by latrines, water sources, or entrance to the house.

Fig 4.13 Tippy tap with foot pedal

How to make a 'tippy tap'

You will need:

- A plastic bottle
- A nail
- A small empty tin can
- A string
- A stick

- A pair of pliers
- A candle
- Matches
- A bar of soap

Fig. 4.14 Tippy tap: Using a bottle with a handle

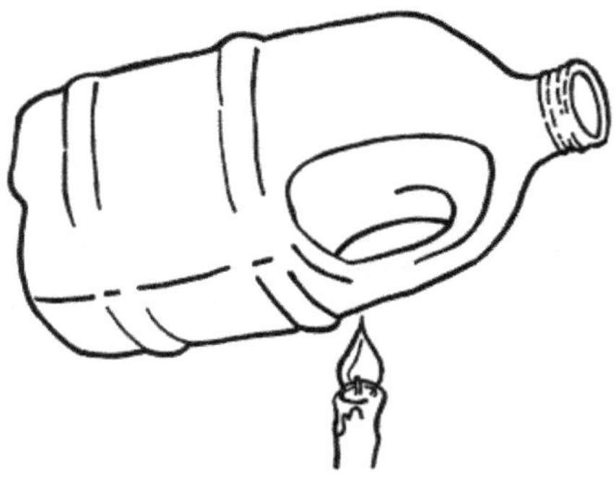

1. Warm the base of the bottle handle over a candle until soft.

2. Pinch the soft part of the handle with pliers to seal it and prevent water flowing through.

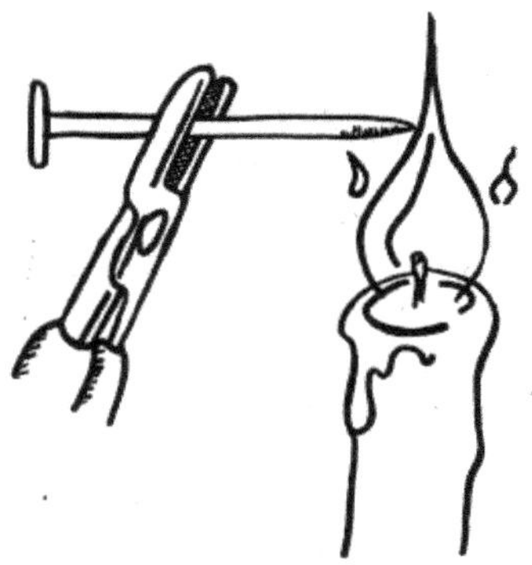

3. Heat a small nail over a candle using pliers.

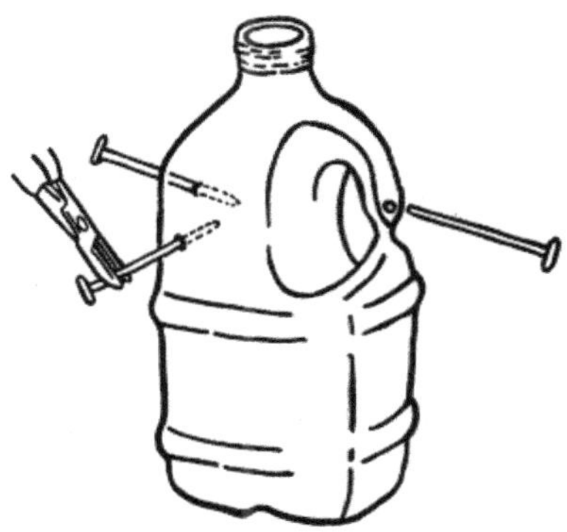

4. Use the hot nail to make a small hole on the outside edge of the handle, just above the sealed area. Reheat it and make two larger holes in the back of the bottle about a thumb-width apart.

5. Thread a string through the two back holes and tie to a stick Pour water into the tippy tap until it is almost level with the holes in the back of the bottle.

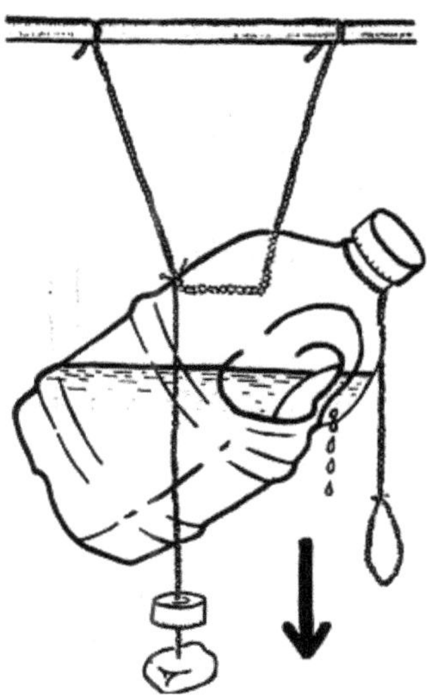

6. Thread another piece of string through a bar of soap and an empty tin can (the base facing upwards to protect the soap from rain and sun. Attach to the tippy-tap string as shown. Tie a third string to the bottle cap and leave end hanging – pulling this string tips the bottle and causes water to flow from the hole in the handle.

Fig.4.15 Tippy tap: Using a bottle without a handle

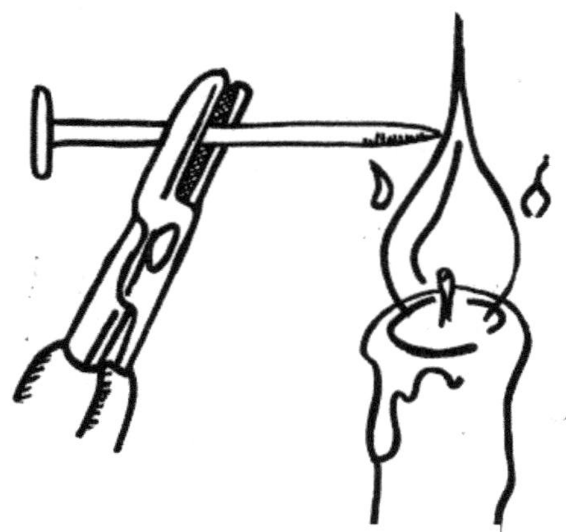

1. Heat the point of a small nail over a candle (make sure that you use pliers to hold the nail).

2. Use the hot nail to make two small holes on each side at the top of the bottle. One hole is for the water to get out and one allows air to get in.

3. Take a long piece of string and tie one end around the neck of the bottle. Tie the other around the body of the bottle, near the base.

4. Pour water into the bottle. S top before the water level reaches the holes.

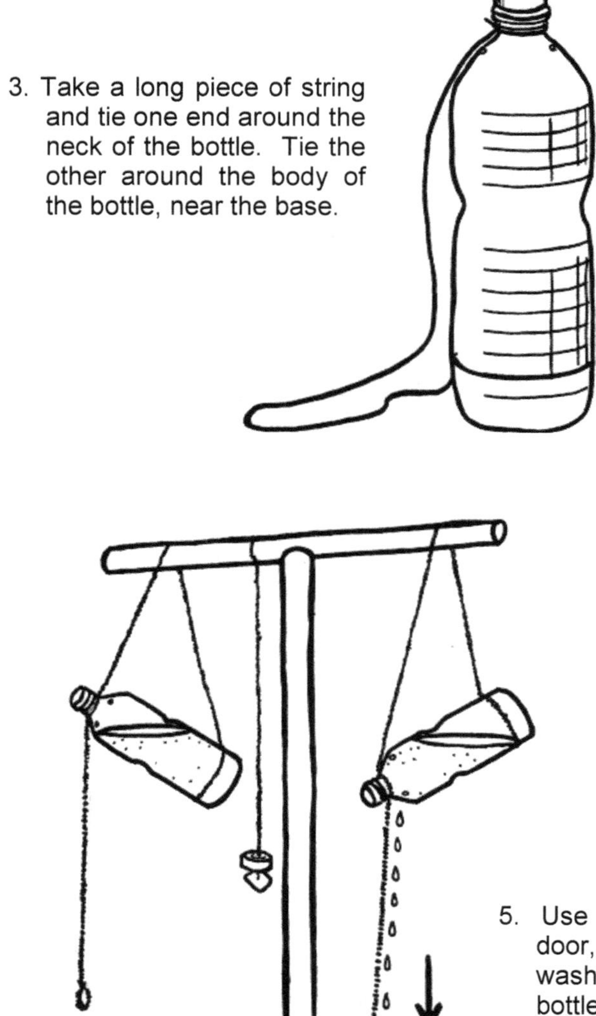

5. Use the string to hang the bottle in front of the latrine door, or where it will best remind the toilet users to wash their hands before leaving the toilet. Tilting the bottle up or down is like opening and c losing a t ap. Hang a piece of soap next to the bottle.

Fig. 4.16 CALABASH HANDWASHER

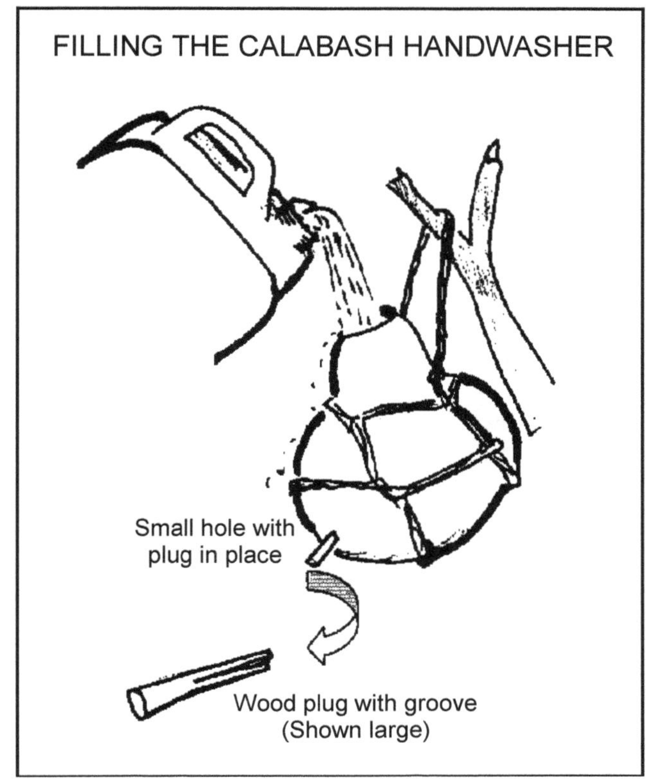

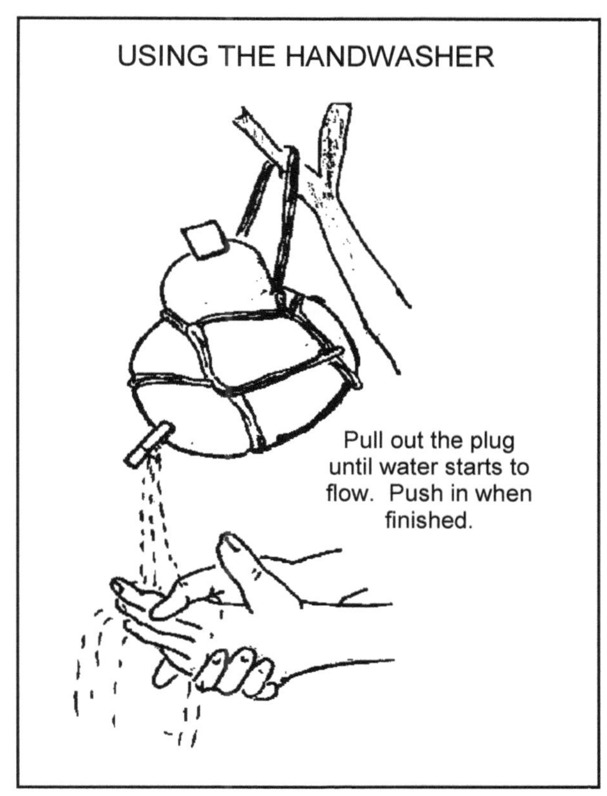

Fig. 4.17 THE LEAKING LADLE

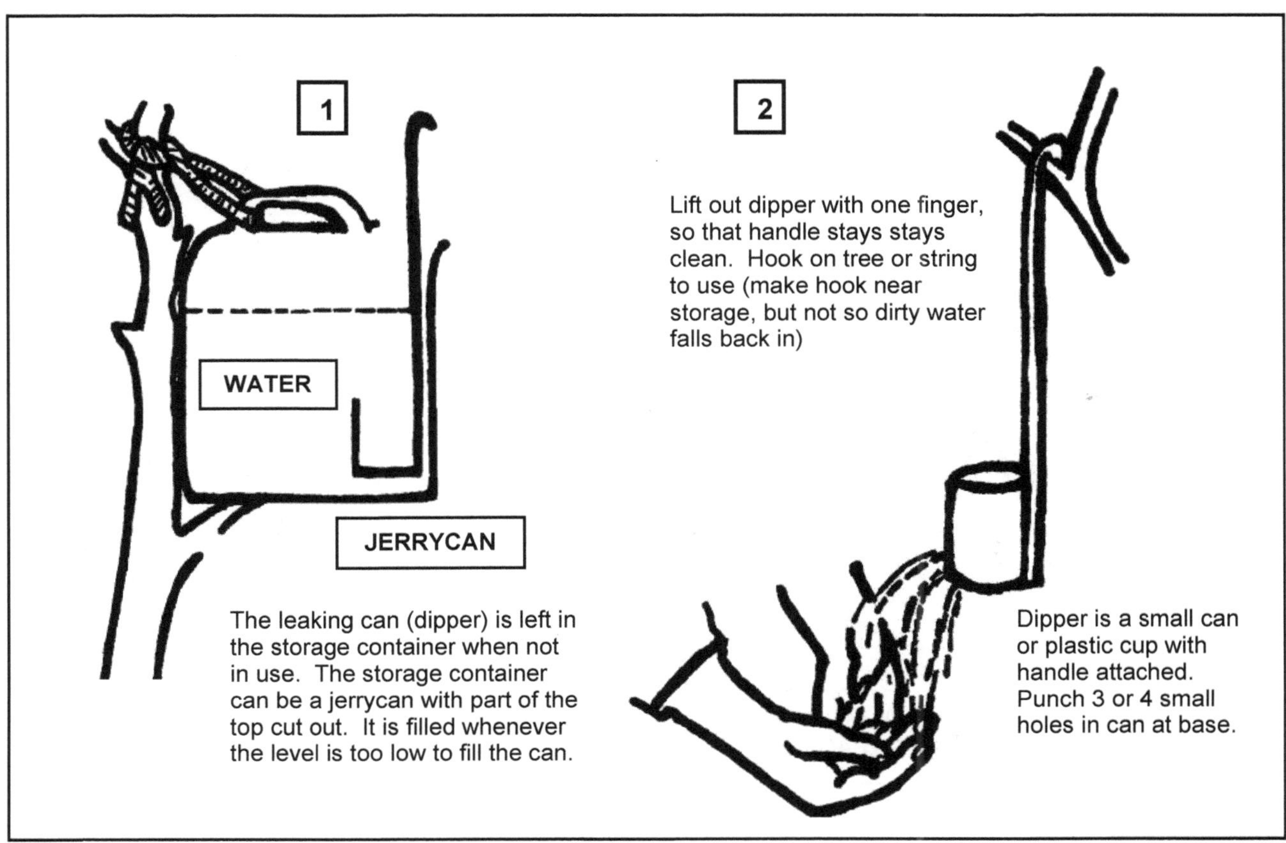

Fig 4.18 Handwashing for houses and schools

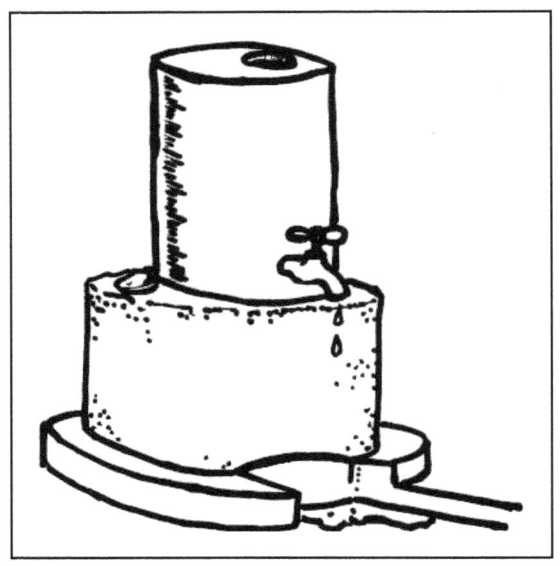 

4.6.2 TIPS ABOUT HANDWASHING

Fig 4.19 How to wash hands thoroughly with little water.

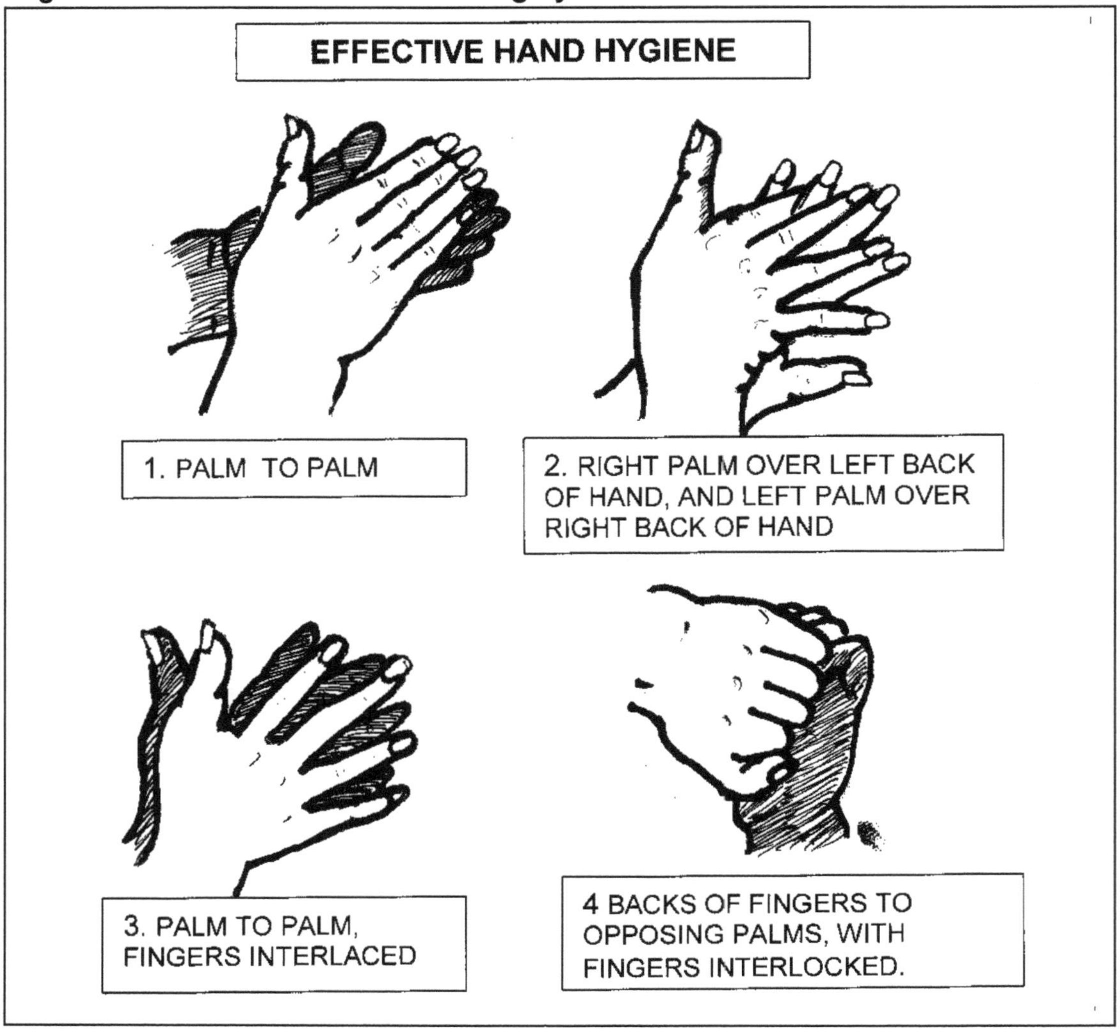

Tips about handwashing (continued)

1. If water is collected when you have dirty hands then the water source will be made dirty. Even if there is only one bucket, this will carry dirt into the water. This means that people need to check that there is water in the house/ in the hand-washer, BEFORE going to the latrine or in the bush. Otherwise they will make the source dirty before they can make their hands clean after defecation.

> **2 scoopholes and 2 lined wells in Western Province always had more than 100FC/100ml in the water. Users did not like this and installed hand-washing devices at each site and made rules for users. After one month the FC count had dropped to zero, throughout each day.**

2. Handwashing can <u>spread</u> disease if not done properly. If people all share the same water in a bowl, the last user may find their hands get dirtier not cleaner! The most effective way is to pour clean water onto hands, and collect the dirty water underneath in a bowl. This can then be used to water plants, or settle dust outside the house.

3. Also when using devices with taps or plugs, these should be handled as little as possible with dirty hands (tip by pulling a string, hook dipper with little finger), or they will spread disease.

4. Soap helps to remove dirt and make people smell clean. Since it also kills many germs it is especially helpful if used after going to the latrine or cleaning a small child/ handling napkins.

5. Children are the worst at washing hands, but also get most dirty and are most likely to get sick as a result, so encouraging them to wash hands and making it easy for them to do so, will have a big effect.... More days at school, fewer trips to health centres and traditional healers, less sleepless nights, less clothes washing (see Module 1.3.)

Fig 4. 20   Placement of handwashing devices: these need to be placed low enough for children to use

> In more than nine out of ten households in four provinces, people said the whole family shared the same bowl of water for washing hands. Very few scooped the water out of the bowl, or poured water to avoid passing their dirt on to others.

4.7 IMPROVING HYGIENE AROUND THE HOUSE.

INTRODUCTION   During community mapping it may have been clear where people feel there are problems of rubbish, scavenging dogs, ponded w ater where mosquitoes breed etc   This session can take these problems and look at them more closely using **the community map** (from Module 1.2) and a set of posters for **'three pile sorting'**.

PURPOSE   To establish standards for environmental sanitation at community and hous ehold level so that risks to health and nuisance to neighbours is minimised.

TIME   2 hours

MATERIALS   Posters of rubbish, food storage, etc., community map, pen and paper.

> Posters used may include:
>
> Selection from **S11-S17**
> **H4-14**
> **G9**

METHOD

1. Explain the purpose of the meeting. Refer back to the community map and problems raised when it was being drawn.

2. Present the set of posters and as k people to describe what each one shows.

3. Ask them to discuss the posters. If there are other features of the surroundings to houses and household practice relating to keeping food clean, discouraging flies and m osquitoes from breeding which they can think of split them into groups and give each a poster to draw to add to the pile.

4. When all posters are ready, ask for them to be s eparated into three piles, those which are regarded as :

    *   Good practice

    *   Bad practice

    *   Neither good nor bad.

5. Ask participants to discuss how people should feel who are seen to follow bad pr actice and w hat measures could be taken at household and c ommunity level to discourage such practices. Do the same with good practices and how people might be enc ouraged to adopt them.

6. Ask participants to conclude whether there are any standards they would like to aim for, towards which individual households could try and pl an so that within a year the situation within the village has changed significantly.

7. Note down what has been agreed by those present as the situation towards which they are working, and how it differs from the present.

8. Summarise, holding up the posters which are relevant. Suggest a meeting after six months to assess what has changed (see Module 6).

DISCUSSION POINTS

*   What are the main problems which affect those within a household?

*   What are the main features which affect neighbouring households?

*   Where households appear to be unable to improve the situation, what assistance can be given or what sanctions could be t aken against those who continue to cause problems for their neighbours, or put their health of their own family members at risk?

*   Suggest another meeting to discuss actions and preferred standards of behaviour and how this can be achieved by all groups, not just those who are progressive or better off. (See Section 5)

NOTE TO FACILITATOR

- Care should be taken to see that there is a distinction made between those who do not have resources to improve the situation for themselves, and those who choose not to do so or are to lazy to do so.

- The old, the disabled, the poor, and those with many children may need assistance. They should be g iven an opportunity to voice their special problems and how they can be helped should form part of any discussions.

Fig 4.21 Clean or dirty? Pigs around the water source

EXAMPLE A village in Eastern province, Zambia, was discussing whether pigs were clean or dirty. The pigs cleaned up after humans, eating rubbish and faeces, but the same could also spread disease. They decided that pigs should be kept away from the houses and water sources. Children's faeces would be put in a latrine, not buried, since when just buried they were still accessible to pigs. Pot racks were also recommended by participants to stop dogs and pigs licking dirty plates. Since pot racks were not an expensive item, those who were able, offered to help the older widows and households with many orphans, to make such racks.

RECORD KEEPING

Note down the main problems identified, and the degree to which each affects neighbours as well as the households concerned. Is there a general desire to improve the situation or do people tend to feel that the present ways are adequate?

NEXT STEPS

Depending on what level the changes will take place (community, household, individual) go to Sections 5.1, 5.3 or 5.4.

4.8 IMPROVING PERSONAL HYGIENE (1)

INTRODUCTION

Personal hygiene is related much to cultural perceptions, but also to availability of water, soap, privacy and availability of alternative clothing. This module looks at what standards people feel they would like to maintain, and how they would do so. There is also a second module looking particularly at hygiene aspects for the target group of teenage girls.

PURPOSE   To establish what is regarded as 'model' condition and behaviour in men, women and children and how this is achieved.

TIME 2 hours

MATERIALS

3 pieces of paper as large as possible, pens, pencil. A lternatively the diagrams may be done on the ground, by drawing around a m an, a w oman and a child.

METHOD

1. Draw figures to represent the body of a man, a woman and a child.

2. Divide people into groups of mixed age but the same sex, and as k a group of women and g irls to consider women and g irl's hygiene, and t he men and boys to consider men and boys'.

3. For each part of the body, get groups to discuss how a c lean person and a dirty person would be. What are the conditions which are desirable and how are these achieved? (What is a 'model' man, woman or child like, and how do they keep like that?)

4. Get each group to present their conclusions and others to comment on them.

5. On the drawings summarise the main points made, and define what a m odel person (man, women and child) would appear to be from the discussions.

DISCUSSION POINTS

- What aspects may affect health, and what are just unpleasant for the person or their friends?

- What local/ traditional and what commercial remedies are there to help reduce problems?

- What taboos exist to re-inforce certain hygiene behaviours, and are they helpful?

PERSONAL HYGIENE

Aspects which in the short term may appear only to be unpleasant (e.g. bad breath or body odour) may in the longer term cause health problems (rotting teeth, skin infections).

RECORD KEEPING

Note down the main problems identified by participants, any related taboos and the solutions suggested by each group.

NEXT STEP

Module 5.4 to establish personal standards and how they can be kept.

4.9   IMPROVING PERSONAL HYGIENE (2)

INTRODUCTION   Dealing with sensitive issues through use of popular theatre helps participants to create awareness through use of song, dance and drama. It assists participants to express their feelings about sensitive issues affecting them and others without making it too personal.

PURPOSE   To establish standards for hygiene at personal level and seek feasible solutions to overcome related fears and bui ld on personal confidence. It can refer to adults or younger people, but is best carried out among those of the same sex and age group.

TIME   2 hours

MATERIALS   Pens, paper, space for performers

METHOD

1. Take a s mall group to develop a pl ay or song on s ensitive issues affecting personal hygiene (body odour, bad smell from the mouth, menstrual hygiene). If there is a l ocal dance or drama group involve them in this.

2. Get them to discuss the issues in detail and c ome up w ith characters in a play or song which depicts an awkward situation which could stimulate discussion in a larger group.

3. Rehearse the play or song in a small group and then invite a larger group of people of the same age to be t he audience.

4. Divide the larger group into boys and girls and as k them to discuss what worries it brings up, what taboos are involved, and what practical solutions can be f ound. Some of the solutions will relate to the ways in which hygiene can be i mproved, while others will relate to how to cope with those of poor hygiene standards.

5. Get participants to identify standards they would like to make 'the norm'.

6. It may be nec essary for groups to present findings separately but there is value in getting the opinion of the opposite sex on t he effects of good and bad practice on relationships, as this is a pow erful tool in changing behaviour!

Fig. 4.22   Peer motivation

GENERAL DISCUSSION POINTS

- Examine what hygiene behaviours are regarded culturally as normal and good practice. Build on those values and see what new ideas people would wish to add on.

- Identify the people whose advice is most respected. How can they assist?

- What problems arise from poor hygiene (these may be social, health, economic, quality of life).

- What standards/ guidelines do participants feel should be s et among the young and adults?

NOTES TO FACILITATOR

Behavioural change takes time. Do not expect instant revolution!

Use role models and respected people such as traditional counsellors and youth leaders.

It may take several sessions before you are ready to make an action plan.

EXAMPLE OF OPEN-ENDED STORY FOR A MIXED GROUP

Kabinda liked the look of the boy. He was tall and well-built, and was doing well in class. However, he seemed to have few friends. She wondered why this should be until she came close to him one day. He certainly made her faint, but not from passion! His breath smelt like rotting leaves and his armpits like very dead fish. She almost choked. She went back to her friends and told them what had happened and what a pity it was that such a good catch should be so unappetising! They started to plot together how they could get Abraham to change his ways ....

NEXT STEP

For action planning go to 5.4.

EXAMPLE OF A STORY WITH A GAP

Chileshe was a pretty young girl. To her amazement each time she walked towards her friends, wanting to mix with them freely, they all ran in different directions. Chileshe continued to wonder why her friends treated her in such a manner.

One day the youth leader from her church gathered enough courage and told her that she produces bad s mells from the mouth and body.

Chileshe wept and w ept and pl eaded with others in order for them to help her find a solution to her problem.

Today Chileshe is a changed person who is so confident and proud of herself. She is now being groomed to become the youth leader because she is so smart, presentable and freely mingles with all her friends.

SPECIFIC DISCUSSION POINTS

1. What was Chileshe's major problem?

2. What could have happened to her if nothing was done about her situation?

3. Does Chileshe's situation happen in real life among young people?

4. What practical measures should be taken or suggested to correct such a situation?

5. What do you think are some of the hindrances to personal hygiene?

6. How can you address these hindrances?

RECORD KEEPING

Make notes on the different problems identified, the physical solutions (such as use of crushed leaves, soap, washing clothes more often) and the social solutions (how to tell someone, the role of elders and peer pressure) and the taboos and beliefs which create barriers or make change easier.

SECTION FIVE

PLANNING HOW CHANGES WILL BE MADE

5. PLANNING HOW CHANGES WILL BE MADE

INTRODUCTION

Having decided what they want to achieve, the community then needs to consider carefully

- **What** actions need to be taken,
- **Who** will carry each one out
- **When** each action will be completed

This requires as many members of the community as possible to be involved and aware of what is decided so that they can support   and encourage those who have tasks to complete. Where changes are mainly behavioural, there will need to be agreement over what will be regarded as signs of success,   and w ho will have responsibility to chase up on those who don't do things in the way which has been agreed.

When a **plan of action** has been made, it can also easily be used to see how things are going, and   whether people are doing the tasks (such as collecting funds or materials) which may need to come before any construction (for example provision of bricks to line a latrine pit or well, collection of funds for cement, or buying of containers to make hand-washing devices).   This monitoring of progress will help the planned changes to be achieved with less delay and fewer problems.

Fig 5.1 Cement, bricks, labour, and equipment all ready on time.

5.1 WHAT ARE WE GOING TO DO? (Community level changes)

INTRODUCTION   Stages 2 and 4 identified the changes that people would like to make. These have been defined by pictures, maps or agreed targets which show how things are and how they would like them to be. These can now be used as a **'story with a gap'** where they will fill in what needs to happen to move from 'how things are' to 'how they would like them to be'.

The extension worker should find out whether any funds are available at district level for the planned changes, before this meeting takes place. He/she should also find out what the procedure is for applying for funds if they are available. Where none are available or people are prepared to provide all skills, labour and materials, then the exercise in Section 3.3 can be used to plan fund-raising, and who will do what.

PURPOSE

To agree in detail what actions are needed when, and who will carry them out. To involve as many people as possible in this stage, so that no-one can say "but we never agreed" or "we never knew it involved all this work", but all can say "it was our idea, and we succeeded". In this example improvement to a communal water supply is used.

TIME 3 hours

MATERIALS   Community map or drawings/posters showing the situation now and in the future. A seasonal calendar if already constructed. Pens, paper.

METHOD

1. Take the 'present' and 'future' pictures/ maps. Ask if anyone has any comments to make on what they show. If any changes need to be made, include them.

2. Divide the participants into two groups. Ask each to discuss what needs to be done to get from one situation to the other. Where possible do this with drawings, but the facilitator should also take careful notes of what is said.

3. After about an hour, (see when discussions are reaching an end) ask each group to tell everyone what they think is needed to be done. When each has finished, discuss any differences and agree final version.

4. If activities are going to take place over several months, place each in the relevant area of the seasonal calendar. Agree at what points meetings may be necessary to discuss progress and any problems which have arisen.

Fig.5.2 Planned Changes: Now and Next Year

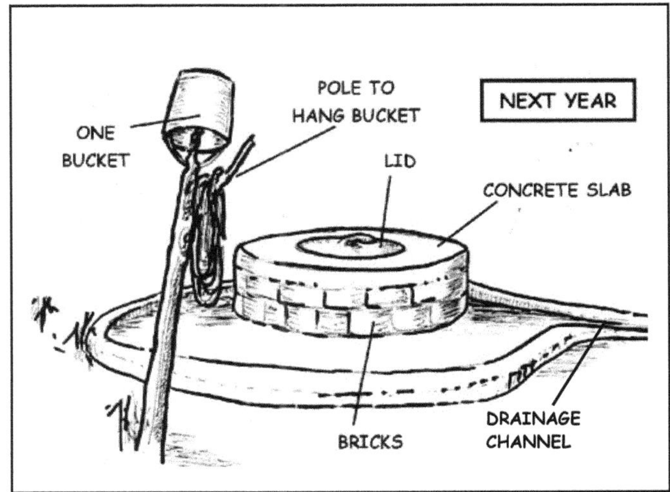

DISCUSSION POINTS

- Is there need t o involve people from outside the community? I f funds are needed from local government, NGOs or district health, or development committees this may take several months to arrange, so needs planning very far in advance. It may require the extension worker to act as a l ink and keep both district and community informed of progress.

- What funds could be made available from within, a) now, b) after the next harvest (see also Stage 3)?

- What are events which may affect the timing of different activities?

- What materials are necessary and how can they be made available?

- What signs will there be that the plan has been successfully completed?

Examples of events

- Initiation ceremony (a good time to spread messages especially on behavioural change)

- Start of rains

- Traders pass through to buy harvest

- Ministry of Health vehicle comes with salaries

- Rising water levels

NOTES TO FACILITATOR:

- This whole exercise is the key to successful changes. It may take more time than you planned, and has therefore been split into two parts, one on w hat needs to be done and w hen and on e on w ho will do each part. For smaller changes 5.1 and 5.2 can be done as one session.

- Agreed steps should be re-assessed and progress checked up on at least every three months.

Example of use.

Improving the water source.

Participants decided that -:

Each household with land will plant an extra row of beans in **August**.

2 bags of beans will be sold in **May** next year to raise funds for cement.

Cement will be brought by Ministry of Health vehicle with **June** salaries to rural health centre. Environmental Health Technician will make sure fibre glass moulds are ready.

In **April**, the freest month for labour, crushing of stone and bringing of sand will start.

Headman will lend ox and scotch cart for one day to bring stone and cement to site.

Meeting in **early June** will fix who will dig out source and how many days work there will be. Work will be done **first week in July**.

When the source is completed, in one year's time, there will be a meeting to decide on how to look after it, how water should be drawn, and any activities not allowed near it. Meanwhile rules made for the present source will be made clearer, and the headman's wife will make sure they are followed.

Indicators

- Water clean and clear, people like the taste

- Carrying containers are not put directly in the water

- Water sufficient for all the year

Table 5.1 Example of a seasonal calendar

Activity	Time	Materials/ needs
Raising funds from beans		
1. Consulting agricultural officer	June	None
2. Planting extra rows	August	Seeds
3. Harvesting and measuring output	April	Pockets/ bags
4. Selling/ exchanging for cement	May	Transport
5. Cleaning out well	July	Picks, buckets, hoes

RECORD KEEPING

Begin to make a table which has activities marked out and the time when they will be done, and what is needed.

5.2 WHO WILL DO WHAT? - TASK ALLOCATION AND TASK SHARING

INTRODUCTION This exercise gives a clear picture of who is going to be responsible for what. It will develop a management structure or work within the structures which already exist in the community. It will also examine the roles of men and women and see where sharing and greater equality of efforts may be possible. It is especially relevant to large tasks such as water source improvement, but parts are also relevant to smaller changes which relate to adoption of standards in sanitation practices, water collection and hygiene, (which are treated separately in 5.3 and 5.4). It will be done by focus group discussion using the plans made in the previous Module 5.1, or in Section 4.

PURPOSE To identify roles and responsibilities, the need f or advice or training, so that plans can be s uccessfully accomplished, and a ny changes be satisfactorily maintained.

TIME 3 hours

MATERIALS Seasonal calendar from 3.3 or 5.1, and matrix from 5.1. Paper, pencils, eraser.

METHOD

1. For a large project, divide participants into two random groups. The first will discuss the tasks and skills necessary to undertake mobilisation of funds and materials and the second the tasks related to c onstruction or improvement. For smaller projects the two can be combined.

2. Take the activities outlined from Module 5.1, which will be s et out on a s easonal calendar. A sk participants in groups to identify both the management and practical skills which will be needed for each activity.

3. If possible, ask people to draw a picture to represent each task. These may include the following :

   - Overall management. The one with the final responsibility to see all goes well
   - Collecting funds
   - Recording meetings and what decisions are made
   - Co-ordinating activities
   - Accumulating and storing materials
   - Working with district and sub-district level authorities
   - Resolving disagreements
   - Organising men

   - Organising women
   - Maintaining standards
   - Dealing with those who do not fulfil their agreed tasks
   - Teaching how to build/ lay bricks
   - Laying bricks/ making concrete rings
   - Woodwork
   - Bailing and cleaning well
   - Making food for those working
   - Bringing materials (wood, crushed stone, sand)
   - Digging pits or well shafts

4. For each task discuss what skills are needed. Divide participants into two random (mixed sex) groups, but make one act as if they are all women and one as men. Get each to discuss which of these tasks could be undertaken by their own group and w hich require assistance from the other. E ach should present their findings and t hen as a whole come to an agreement as to how to share the tasks without putting too much burden on one or other group. Identify by name, who would be best able to undertake each task. Are those people present and are they willing to take on the responsibility?

5. Ask participants to identify whether there are any training needs for skills not available within the community, or whether it is better to pay others to do those tasks (especially if training is not easily available). What advice can be provided by extension workers and district officers?

6. You should end up with two outputs.

   - One is a c lear picture of the management structure for the work.

This may be a committee or it may be a traditional or extended family leadership, depending on the views of the participants.

- The second is a c hart which records what will be done w hen, by whom and what are the necessary materials, skills, advice or funds needed for that activity (see Fig 5.3).

DISCUSSION POINTS

- Is the traditional structure or a new group (which may include retired civil servants, pupils just finished secondary school, women with organisational skills who are widely respected), a better system for managing change? Or can the two be successfully combined?

- Planning and attending meetings is time consuming. H ow can plans be made and progress monitored without taking up too much time among a large group of people?

- What tasks might better be done by raising funds and using a local contractor (eg. Deepening of a well)?

- How can the poorest be i nvolved without putting undue burden upon them?

- Is there a need for a management group to cover all aspects of sanitation, and environmental hygiene as well as water?

NOTE TO THE FACILITATOR:

Try to provide a structure to discussions so they do not wander off into unnecessary detail, but do not lead the discussion and impose your ideas.

Offer your support and advice if asked for, but encourage the community to use its own skills as far as possible. By the time the work is done y ou may be posted somewhere else, and then, if the project is too dependent on your inputs, it will fail. This is also a g ood reason to keep good records, so that if you move away, someone else can continue your good work.

RECORD KEEPING
Note down the names of those who will manage the work. Make a table (matrix) of what will be done (actions), when (timing), materials needed (requirements) and by whom.

Table 5.2 EXAMPLE OF ONE ACTIVITY AND ASSOCIATED TASKS

WHAT	WHEN	WHO	REQUIREMENTS
Providing crushed stone	**April/ May**	**Thomas Mubanga**	**Cold chisels, hammers Transport**
Bringing large stones to the site	April	Thomas Mubanga	Scotch-cart/ ox/ two assistants. Skill: managing animals
Providing cold chisels / hammers	April/ May	Danny Kawama, Dominic Mwila,	Cold chisels, hammers
Crushing stone	April/ May	Maselino Bwali, Chilufya Mulenga, Faustina Chisanga, Dominic Mwila, Thomas Mubanga	
Measuring amount produced	April/ May	Thomas Mubanga	Bucket Skill: Calculating, recording amounts

5.3　WHAT ARE WE GOING TO DO? (Household facilities)

INTRODUCTION This module looks at changes at household level, and i s equally relevant to increasing availability of water in the house (Module 4.5), hand-washing (4.6) and household hygiene (4.7) as well as sanitation (4.3).   Sanitation is taken as the example. Building or improving a l atrine is usually the responsibility of the household head.   Problems often arise because of poor practice in disposal of faeces, especially among children. People were therefore asked to go and discuss with their families what they had found out from Module 4.3 and what sanitary practices they would like to change. This module uses the posters of the preferred options from Module 4.3 and group discussions, but can equally use the outputs from 4.5-4.7 in planning which options to follow.

Fig.5.3 Teaching children to wash their hands

PURPOSE To establish community norms of behaviour in relation to defecation or hygiene and how they will be enc ouraged/ maintained. Each household will plan what they feel they could achieve within a year. These changes   may include up-grading or new construction, as well as aspects such as disposal of children's faeces, or adoption of cat method or sharing of facilities by those without latrines. Equally they may involve provision of hand washing jugs or devices, or making rubbish pits and plate racks.

TIME   1 ½ hours

MATERIALS

Posters showing different (sanitation) options; plastic folders; counters / stones; paper and m arker / drawing pens.

TARGET G ROUP   Representatives of as many households as possible, and village elders/ influential people.

METHOD

1. Remind participants of t he main points made Module 4.3 (or 4.5-4.7)

2. Set out the posters of sanitation options. Ask each household which is represented to put a stone on the option they plan to be using by the same time next year.

3. Divide participants according to the number of alternatives selected. Ask each group to discuss what materials and skills would be needed to construct or improve facilities to achieve the alternatives selected.

4. Ask each group also to discuss what changes they feel they, as a family, can make to their present practices and how this may benefit both themselves and the community as a whole. A sk them also to discuss how they might be able to assist those who are unable to make changes (eg. Those without latrines, assisting them with construction, or providing materials or by sharing). Get each group to present its conclusions, and then work out what materials may be needed to be purchased and how payment can be made if this is not being done a t household level.

5. Summarise the findings and t he 'community guidelines of acceptable behaviour' which come out of the discussion and obtain agreement on what to do if some families do not act according to the guidelines.

DISCUSSION POINTS

- What can those not able or prepared to construct a l atrine offer to do to improve the situation and w hat could others do to assist them?

- Summarise the conclusions relating to community plans and what practices they feel are acceptable and not acceptable. Make a check-list of which households are planning to build latrines of different types, (or to adopt different methods of hand washing) and what changes others are offering to make.

- Agree what will be the indicators of change which will be u sed after a year to judge how successfully the targets have been r eached. (for example these may be specific such as who has access to different types of latrine and us es them regularly, or general, such as 'no faeces found lying around within 75m of the houses'.)

- Discuss who within the village will be the ones to monitor progress, and encourage people to complete what they planned.

- Outline the next steps and agr ee next meeting.

NOTE TO FACILITATOR:

It is important that the session is used to agree specific sanitation changes that different households, families and individuals intend to make. All members of the group should be enc ouraged in the choices they make based on t heir resources and pr iorities, and t hose families opting for practices or latrine types at the 'lower' end of the Sanitation Ladder **must not** be criticised for their choice. 'Something is better than nothing' or 'step by step' should be t he motto of the session.

However this is **not** to say that community's decisions cannot be explored and i nvestigated. If possible, the reasons for people's choices should be probed as it is important that there is clarity on the motivation for people to invest in new sanitary facilities. This information will be extremely useful in guiding future efforts at promoting sanitation in communities

Community members at Choobana, Monze agreed that:

- Each household would confirm their choice of the latrine they were to build during house to house visits by the extension worker. The extension worker would help calculate the number of pockets of cement that would be needed. If funds for this might be available from any NGO or local water and sanitation committee, the extension worker could send in a request for these.

- The extension worker would conduct a practical builders training course for 12 people chosen from within the village. This course would focus on how to build the types of latrines that the members of the community wished to construct.

- Each household would then collect the local materials needed to build their latrine and dig the pit to the dimensions required. Each household would also dig a small pit in which to put their rubbish and construct a rack on which to put plates and pots to dry.

- People also agreed to start covering any left over food to stop flies from sitting on it.

- Groups of three households were to be formed to support each other in the various tasks and to monitor progress. Each group of three households would also monitor and support their neighbouring group of three

- Progress on all of these actions was to be reported at a meeting with the extension officer after a month when the number of pockets of cement to be delivered would also be confirmed.

- The cement would then be delivered and the trained builders would construct their own latrine first and then others for the rest of the households in the community. No payment was to be made to these builders for this work.

- After another month the extension worker would meet with the community again to review progress and identify any problems that were preventing people from completing the work.

Indicators

- All households have dug a pit to the right dimensions and have collected the sand, stones, poles and grass needed to build their toilet.
- Latrine builders have been trained and have constructed and are using their own toilet. Each builder has built at least one other latrine in the village.
- All households opting for latrines (of any type) have built and are using their toilet.
- There are no human faeces found lying around within the boundaries of the village.

<table>
<tr><td colspan="2">NEXT STEPS</td></tr>
<tr><td>If other changes in handwashing, personal or household hygiene are also associated by participants with changes in sanitation</td><td>Go to Module 5.4</td></tr>
<tr><td>If no other changes are considered necessary</td><td>Get started without delay!</td></tr>
</table>

RECORD KEEPING

Note down the changes each house wishes to make. You can use these notes on later visits to discuss progress with individual households and t he community as a w hole. N ote also material and training needs which may require outside assistance.

Example of use

Improved and more frequent hand-washing practice

Actions proposed by participants and agreed.

Within one week. The headman to ask local trader to look for second hand 2.5 litre Coca-Cola bottles and preferred soap types, at good price.

Each household head will see that family do not have to share water from the same bowl (a special pouring cup is there) and if possible that soap (or substitute) is available for all to use

Women to show children how to wash hands well, without wasting water, and discuss with household head is storage containers are sufficient.

Environmental Health Technician shows women how to make hand-washing devices for latrine or house entrance.

Follow-up meeting after two months to see if any changes have happened and if people like the changes.

Indicators

- At least half the houses have facility (cup and bowl, or device) for those entering the house.
- All families pour water for hand-washing, do not share bowl, before eating.
- All children especially, have clean hands before eating.

5.4 WHAT ARE WE GOING TO DO ? (Behavioural change)

INTRODUCTION To change behaviour at an individual level generally requires not just the knowledge of what good behaviour is, but also pressure brought about by the opinion of others. 'Others' may be friends of the same age, or elders whose views are respected, either within the family or outside. This module refers particularly to personal hygiene and behaviour around the water source, but can be applied to other areas of behavioural change. The module uses focus group discussion and role play, plus, body mapping from 4.8 for personal hygiene.

PURPOSE To establish behavioural norms, and identify the pressures which exist which can help in their maintenance, so that personal hygiene (or household hygiene or behaviour at the source) improves through a combination of personal wish and peer group pressure. In this case personal hygiene is used as the example.

TIME 2 hours

MATERIALS Outputs from Module 4.8 and /or 4.9 or 4.4 or 4.6

TARGET GROUPS

Teenagers and adults. For personal hygiene meetings would preferably held with single sex groups, but including different ages.

METHOD

1. Divide participants into two groups, adults and teenagers.

2. Summarise the views which came out of the Modules 4.8 and / or 4.9 (or 4.4). What are the chief standards that people feel are important in personal hygiene for those growing into adulthood? Get each group to discuss them and p rioritise which are the most important. If possible do this using drawings or a body maps (from 4.8) for each group.

3. With the standards and priorities agreed ask each group to discuss how these standards could be encouraged, and in what circumstances can they be eas ily promoted?

4. Ask each group to discuss how they would encourage the keeping of standards and how they can best let someone know when they appear to have poor personal hygiene. Get each group to act out one ex ample of someone with hygiene problems and how they would help t hat person to overcome them.

DISCUSSION POINTS

* Initiation ceremonies may be used to re-inforce messages to girls as they reach puberty, but what opportunities are there to educate and inform boys and who should be responsible for this?

* What indicators should be taken of good hygiene, and what standards set?

* How can peer groups (people of the same age) best let someone know that they smell, without causing offence? Does this differ for elders and teenagers?

* Can 'model' individuals be used as a way of encouraging change? If so who are they, and what indicators could be us ed to show their good practice?

* Who is ultimately responsible if someone fails to maintain reasonable standards and what sanctions do peopl e feel could be applied?

> **NB.** The same questions could be asked of practices near the water source, or household hygiene.

EXAMPLE ON PERSONAL HYGIENE

Participants decided to concentrate on oral hygiene and stopping the smell of bad breath as a first priority. This would also reduce tooth-ache and stop people's teeth falling out so early. Each would have a stick or brush to clean teeth and use leaves or soap to help get teeth clean. At other times of day they would rinse their mouth after eating, and avoid chewing so much sugar cane.

Responsibilities: Parents would check if children had cleaned their teeth at least once in a day and would complain if their partner or any member of the family had bad breath. Those who did not manage to avoid bad breath would be taken to the herbalist who would try and assist, but this would cost money, so people should make an effort themselves to solve the problem.

Indicators: Less complaints of bad breath, less toothache.

EXAMPLE RELATING TO KEEPING THE SOURCE CLEAN

At Kazuali, users of the source decided that everyone should use the handwashing device before drawing water. They put a mark on the bottle to indicate when the level was low. The next user was to fill the bottle after handwashing. Children were not to draw water and no napkins were to be washed within 50m of the source. The source area was to be kept swept, no rubbish was to be put within 50m and grass was to be cut low. Water was not to pond and the soakaway was to be kept clear. When it was blocked, the chairman was to be informed.

Responsibilities:
Headman's wife to make rota for cleaning around the source. Each family to inform their children and adults of the rules, and household head to enforce good behaviour.

Users to take it in turn to provide new handwashing bottle and soap as needed, headman to organise.

Senior wife and headman to be final judges if bad behaviour continued

Men to cut grass and clear soak-away on rota organised by headman

Indicators:

Handwashing bottle always with some water in it, but used

Source area clean and low grass, no water ponded nearby

Children not seen playing near source, nor napkin washing observed

SECTION SIX

MONITORING AND EVALUATION OF PROJECT IMPLEMENTATION

6.1 MONITORING PROGRESS.

6.2 ASSESSING ACHIEVEMENTS

6.3 EVALUATING CHANGES

6.4 EVALUATING THE IMPACT

6.4.1 HEALTH IMPACT

6.4.2 IMPACT ON THE COMMUNITY

6. IMPLEMENTATION

INTRODUCTION

Details of how to construct or improve facilities is given in the associated guidelines on ' Source Improvement' and 'Sanitation Improvement'.   During implementation of plans made with Module 5.1-4, (whether they relate to changes in behaviour or up-graded facilities) commitments will have been made by people to undertake specific tasks and funds and other resources may have to have been collected. Many planned activities will only be pos sible if the previous steps have been c ompleted. At various stages therefore, it will be necessary to check if there are any delays, and how they can be overcome. Some may mean that original plans will have to be modified. The frequency of meetings to assess progress and help solve unforeseen problems will have been decided in Module 5.1-5.4 but can be r e-assessed as implementation proceeds.

Generally the information gained from monitoring progress will also feed into evaluating the successes and failures of the planned changes.

Monitoring progress is a procedure which works through a cycle.

- Collecting information,
- identifying bottle-necks or problems which require design to be changed,
- re-scheduling later activities, and then after a period
- collecting information again t o s ee how things are going.

In this way a pr oject can be s uccessfully completed with as few hold-ups as possible, and s o with minimum confusion or disagreements. Where contractors are involved it will also reduce costs.

6.1 MONITORING PROGRESS

INTRODUCTION   Those who have been put in charge of managing planned changes within the community (often a N eighbourhood Health Management Board, traditional leaders (male and female) or Community development or local water and sanitation committee) will need to encourage all those involved. This they can partly do by showing what progress has been made, and also by assisting those who find they are having difficulties to complete what they agreed to do. This is done mainly as focal group discussion using the calendar developed in 5.1 or targets identified in 5.3 or 5.4.

PURPOSE   This session should be carried out at regular intervals. It is designed to inform participants of how things are progressing and to avoid problems which are developing.

MATERIALS   Project calendar from 5.1 and indicators (especially for behavioural change).

TIME   1 hour

METHOD

1. Take the plans developed in 5.1 or 5.3 or 5.4. Ask participants to indicate what stage has been reached.

2. Ask them to indicate what has **not** been achieved which is needed before the project is complete and to come up with reasons for this.

3. Discuss   whether the experience so far means that the original plan needs modifying. If so, draw up a new calendar, or mark changes to the old one.

4. Agree   immediate   actions   to remove bottlenecks and when the next monitoring meeting should be held.

DISCUSSION POINTS

*   Have any of the people who were to be involved left the village, so that their role needs to be taken up by others?

*   Have   funds   not   been   made available? If not, should the plans be changed in terms of **the product that was planned** (eg. use a lower technology,   change   materials, provide more labour from within the community,   rather   than   by contractor) or in terms of timing (should   certain   activities   be delayed)?

*   If things are going better than planned, does this mean that more could be achieved, and plans should aim higher?

NEXT STEP
Continued monitoring until project completion, and then evaluation of the changes and their impact. (Section 6)

6.2 ASSESSING ACHIEVEMENTS

How did we do?

INTRODUCTION

As a result of the previous steps you have undertaken from this manual, changes may have taken place in behaviour, and/ or in water supply and sanitation facilities. To learn what is effective and what makes little difference, and what is liked and not liked, it is necessary to find out:

- **what changes have taken pl ace** directly as a result of the steps taken,

- the **effects such changes** have had, and

- **how** different groups of **people view such changes**.

You will then be able to answer   the two questions -:

How else could changes have been made?

Are there are better ways to achieve the same results, so that the efforts you and others make are as effective as possible? The changes people see and the reasons they like or dislike them will help you (and them) to find good ways to encourage others to change their way of life so that it becomes more healthy and enjoyable.

And

 What still needs to improve?

Looking at what has changed, and what remains to be done, and relating it to the previous discussions, will help communities to see what other aspects they could improve upon.

6.3 EVALUATING CHANGES: What did we change? (SHORT TERM CHANGES)

INTRODUCTION   Where changes have been made to water sources or sanitary facilities, these can be l ooked at alongside any changes in behaviour with which they may be as sociated. Otherwise the target behavioural changes and indicators defined in Section 6.1, can be us ed on t heir own as the starting point for discussions.

PURPOSE   To see whether the planned changes were successfully made, and what were   t heir effects. Some of these effects may not have been recognised during planning, and there may be unwanted effects as well as the desired ones.

TIME   2 hours

MATERIALS   Any posters or drawings/community maps which were used in **Action Planning** which were used to define the 'before' and the expected 'after' situations).

METHOD

1. Divide participants into groups of men and women. S et up t he posters/ drawings where both groups can see them (or better still, copy them by hand before the meeting). Review, using the dr awings, what changes were wanted and why.

2. Ask each group to discuss the changes that have taken place. Are these the changes that were expected / planned? Or would they like to re-draw the 'after' picture(s)?

3. Note any differences between the views of men and women.

4. Summarise the main points made, and any actions people still feel need to be t aken, or should be planned to follow on as the next step in improving quality of life for participants and other members of the community.

DISCUSSION POINTS

* What changes have taken place?

* What lessons have been l earnt, both about working together to make changes and al so on technical aspects.

* What could have been done better if it had be en done differently?

* What good things came out of the activities and pl anning, which could help others to do the same?

* Using the community map, were there different views or degrees of involvement of different households/ areas in the village?

* If changes are small and only one session is planned to discuss them, take each element and as k how it has changed people's lives.

* Are there differences for men, women and children?

Fig. 6.1 Easy access to a good water source frees women's time.

RESULTS OF THE EVALUATION

FIG 6.2 What changes were planned and what was achieved?

BEFORE   **PLANNED**

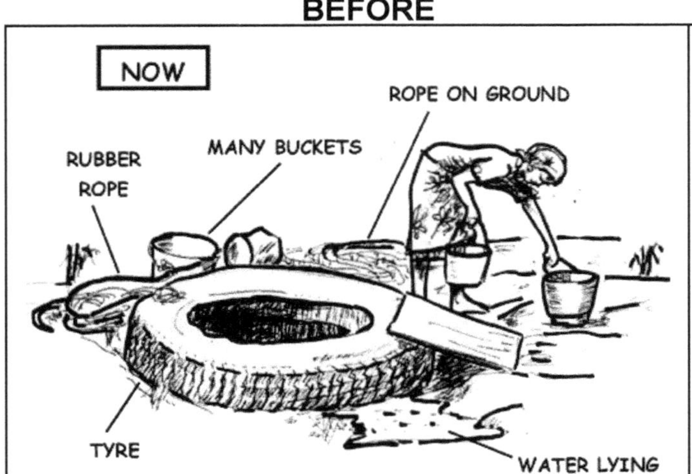

AS COMPLETED

RESULTS OF EVALUATION: What they said

The water is now clean.

Now we drink from the source near the house, and don't take time to go to the spring which is far.

We women asked for a low brick surround to keep out dirt. The men made a high one out of concrete rings. It is difficult to lift water when we cannot stand over the opening.

We are glad that children are no longer in danger, nor can dogs fall in.

Our neighbours have put brick pillars and a windlass. They have moved vegetable growing nearer, and made bricks for their house. We want to do the same.

The men don't want to break up the slab to put in posts and say the hole in the top is too small. They say we need to collect more funds to do this. We plan to change it after next harvest.

We are proud that we made the changes and will know better how to plan next time. Men and women worked together on this.

Children are pleased that they do not have to collect drinking water, and that the surround is too high for them to draw water.

6.4 WHAT EFFECTS HAVE THE CHANGES HAD? (LONG TERM CHANGE)

INTRODUCTION The first section has looked at the direct changes which were made and how well they have been done. However those changes were made more for their **effects** not just for the sake of change. These effects may be on wider aspects of quality of life, and some may have been achieved, whilst others may not, and yet others which were not expected may have occurred.

This section can form part of the same session but may, especially where changes have been great and some time has passed since they were made, require more in-depth discussion. Participants may have their own ideas of what should be discussed but the following two tools are suggested to provide a framework.

1. Body mapping

2. Group discussion on livelihoods, health and hygiene, and socio-cultural aspects which may have been affected over time.

6.4.1 BODY MAPPING FOR HEALTH IMPACTS

INTRODUCTION This explores both changes in hygiene and also effects on the body of changes which have been implemented. In the case of water supply this may include those relating to increased amounts of and decreased distance to water, but also new methods of water lifting and transport. It should explore both positive and negative changes, relating to health and well-being.

PURPOSE To identify with the community the physical changes resulting from changes in water supply, hygiene and/or sanitation.

TIME 1½ hours

MATERIALS
Large sheet of paper, marker and pens.

METHOD

1. Ask a participant to draw the outline of a child (or if enough paper, draw around a child, or around hands and feet, and face. This can be a good ice-breaker). The outline is considered to represent all conditions... child, adult male, pregnant female.

2. Discuss whether anyone has seen or experienced changes, by considering aspects as in Table 6.1.

3. Divide people into small groups and get each to discuss any changes to two or three of the aspects listed above. Groups may be of elderly people, men, women and children, or whatever grouping people feel is appropriate.

4. Ask each group to present the results of its discussion and write (or ask someone else to write) the results on the body map.

DISCUSSION POINTS

- What was the situation on personal cleanliness before?

- What has changed?

- What areas require further improvement?

- Are there changes to the physical / health condition of any aspect discussed (eg. incidence of disease, relating to each area, back-ache, muscle strain, tiredness etc.)?

- Have these changes had any effect on other aspects of life? (eg. absence from school, trips to health centre, spending on treatment, general well-being, time spent on looking after the sick). These are explored more in the next section.

**Table 6.1   IMPROVEMENTS IN HEALTH -
EXAMPLE FOLLOWING ON HYGIENE AND NEW WATER SUPPLY**

BEFORE	ASPECT	AFTER
Scalp itchy, skin flaky, headaches from carrying water far	Hair/head	Fewer headaches, scalp less itchy
Flies and 'glue' around children's eyes	Eyes	Better cleaning of children
Noses runny among children. Dust made people sneeze and wheeze	Nose	
Breath smells and teeth 'sticky'	Teeth	
Ache from water carrying, and drawing water by hand	Neck and shoulders	Windlass reduces strain, more carrying by bicycle
Smells from armpits and other areas	Body	Use more water and more soap, feel fresher
Clothes not clean, difficult to dry near source	Clothes	Can wash clothes at home, without leaving children, and hang by house.
Smell, cuts go bad from mud around source	Legs/ feet	Easy to get at water cleanly.
	Skin	
Under 5's frequent diarrhoea and in elderly. Difficult to keep them clean	Bodily functions/digestion	Easier to clean up and under 5's less often sick.
	Hands	
Tiredness at the end of the day, washing clothes and c hildren takes time and o ften have to wait for water.	General well-being	Less queuing and t iredness, more time to look after children, less sickness in the house. C an give water to livestock easily, and w atch children while fetching water. Use more soap but soap expensive

6.4.2 COMMUNITY IMPACT ASSESSMENT

INTRODUCTION Using the 'before' and 'after' situations as a basis to analyse the main effects the changes have had on way of life forms the core of this session. However since some changes may be independent of the program/ or project (eg increased consumption of soap or ownership of jerrycans as economy improves), a similar exercise should also be done in 'control' communities (those where no planned changes were undertaken), if the effects of the project are to be isolated. In this case participants would identify and draw the situation five years ago and now, or whatever is the relevant time interval.

Fig.6.3 Community using improved water source

PURPOSE To show the expected and unexpected impacts of particular interventions (changes), and how these vary among different groups of people. B oth positive and negative impacts should be recorded.

MATERIALS Large sheet of paper or several A4 sheets. Markers/ pens

TIME 2-3 hours

GROUPS Elderly, children under ten, men and women, young men and girls (if insufficient numbers, consider impacts on eac h group within overall discussion in larger groups). In the later part of the discussion give those not involved in the changes an opportunity to speak as a group or as individuals.

METHOD

1. Ask participants to review the findings from Section 6.3 and i f used, also Section 2. I f necessary add det ails from your own notes.

2. Then ask them to consider their daily activities, production of food and income, hom ecare, beliefs and relationships within the household and in the community as a w hole. What is the same, what has changed? (Posters G.1 to G.14, and W.26, Daily activities) can be used to assist in this.)

3. Write, or get participants to draw pictures of, the main i mpacts on separate pieces of paper, which can then be gr ouped under headings or as positives and negatives.

4. Summarise the findings under headings such as

   - use of time,
   - livelihoods/economic,
   - social/cultural,
   - health,
   - education,
   - quality of life, community well-being.

EXAMPLES FROM WATERAID MONZE

Impact of new water supplies on livelihoods:
- Shortened distances mean time for basket weaving, pottery making, charcoal burning
- Growing and selling more vegetables
- More / healthier goats and more milk to sell
- Moulding bricks to build own houses and to sell
- Rearing pigs and more chickens
- More cash, so can pay clinic fees

Impact on health & hygiene:
- More cooked meals and more food (more time to prepare)
- Children drink more goat's milk
- Regular bathing for children and adults
- Fewer body sores
- Regular washing of clothes
- No infected / cracked nipples during breastfeeding
- Decrease in diarrhoeal diseases
- Clean utensils and plates
- Some latrines have been constructed
- No back-ache / body pain from carrying water for long distances
- No headaches from carrying water on the head
- Reduced visits to clinic
- Bigger / better houses

Impact on well-being:
- Drawing water at any time of day with no fear of abuse
- No need to provide labour for water
- No quarrels with husband over late return and delayed cooking
- Use of "our own" water point
- Husbands now love their wives because they are clean and both don't smell (improved relations)
- No waking up early to collect water
- Shorter distance, less tiredness , more time

Impact on children / education:
- We are able to pay school fees for our children
- Children go to school on time (not out collecting water, or watering goats)
- More time to attend to children
- Purchase school books from sale of goat's milk
- Parents not summoned to school for children's late attendance

DISCUSSION POINTS

- Which impacts were expected and which were not recognised during planning and making the changes?

- Could negative impacts have been avoided? If so, how?

- If there are participants who were not involved in the changes (by choice or by being left out) ask them for their feelings on these changes and on their own way of life.

- Do those who were involved (stakeholders) feel that the efforts they made (costs in terms of time, energy, funds, materials or whatever their contribution) were worthwhile?

- Do those who were not involved by choice now feel they would like to have acted differently or not?

- Of the impacts identified, what do participants feel were the ones which made their efforts most worthwhile?

- Do the successes make the participants want to tackle another problem? If so, what, and where should you begin in the process you have just completed!

References / additional reading material

1. Rural Water Supplies and Sanitation. Peter Morgan. p 359. MacMillan 1990

2. Hand-dug wells and their construction. S.B. Watt and W.E. Wood. p 253. IT Publications 1979

3. Environmental Health Engineering in the Tropics. Sandy Cairncross , Richard Feachem. p.306 Wiley 1996 *(Available from TALC)*

4. Studying hygiene behaviour. Sandy Cairncross , Vijay Kochar. p 334. Sage 1994

5. Water Quality Monitoring. Jamie Bartram and Richard Balance. P 383 E& FN Spon. 1996

6. Partners in Evaluation. Marie-Therese Feuerstein. TALC/ MacMillan p196 1986

7. Tools for Community Participation Lyra Srinivasan. p 179. UNDP 1990.

8. PHAST Step by Step Guide. Sara Wood, Ron Sawyer, Mayling Simpson-Hebert. p 127. WHO 1998

9. Hygiene Promotion. Suzanne Ferron, Joy Morgan and Marion O' Reilly. p250. IT Publications/ CARE International 2000 *(Available from TALC)*

10. Latrine Building: A Handbook for Implementation of the Sanplat System. Intermediate Technology Publications, London 1997, with support from Sida and UNICEF

11. Ecological Toilets. (Pre-production draft), Peter Morgan, Stockholm Environment Institute 2009

12. A Community Guide to Environmental Health, Jeff Conant & Pam Fadem, Hesperian 2008 *(Available from TALC)*

13. Low Cost Water Source Improvements: Practical Guidelines for Fieldworkers, Sally Sutton, TALC, 2004 *(Available from TALC)*

1. Well-digging and construction:

Reference (book or manual)	Available from:
Hand-dug wells and their construction S.B Watt , W.E Wood IT 1998	www.practicalactionpublishing.org www.developmentbookshop.com
Upgrading Well Manual for Field workers Mvuramanzi Trust 1995	www.mvuramanzi.org.zw
Low Cost Water Source Improvements. Practical Guidelines for Fieldworkers S. Sutton 2004	www.talcuk.co.uk (downloadable)

2. Rain water harvesting:

Domestic rainwater harvesting. WELL factsheet Jo Smet 2003	www.lboro.ac.uk/well/resources/fact-sheets/fact-sheets-htm/drh.htm
Domestic roofwater harvesting research programme. Warwick University. (designs, pilots, case studies)	http://www.eng.warwick.ac.uk/dtu/rwh/index.html
Texas Guide to Rainwater Harvesting, Texas Water Development Board 1997	http://www.eng.warwick.ac.uk/ircsa/factsheets/TexasRainHarv.pdf

3. Low cost pumps and other mechanisms:

Canzee pump (low to medium lift, direct action)	www.swsfilt.co.uk/weblog/2007/07/canzee-pump-news-summer-2007.html www.bushproof.biosandfilter.org/assets/files/Canzee_brochure_version_240506.pdf
Rope pump, + wind, high lift and other versions	www.ropepumps.org Spanish and English versions www.ropepump.com
Description of principles for each type of low cost pump WEDC	http://www.lboro.ac.uk/well/resources/technical-briefs/35-low-lift-irrigation-pumps.pdf

4. Social and financial aspects - supply ownership and sharing financial mechanisms for small scale investment:

Microfinance for water supply services, C Fonseca 2006 WELL Factsheet	http://www.lboro.ac.uk/well/resources/fact-sheets/fact-sheets-htm/Micro%20for%20water.htm
Assessing microfinance for water and sanitation. Meera Mehta July 2008 B and M Gates Foundation	http://www.gatesfoundation.org/learning/Documents/assessing-microfinance-wsh-2008.pdf
Microfinance for Sanitation D Saywell/ C Fonseca 2006 Well factsheet	http://www.lboro.ac.uk/orgs/well/resources/fact-sheets/fact-sheets-htm/mcfs.htm

Please adapt, please copy and please give us feedback!

This manual describes projects and case studies which we hope will prove useful and might help with the improvement of water supply and sanitation provision. However, we are aware that there are gaps in our experience and understanding, especially in the way these solutions may be applied locally. What is a good solution to a problem in one area may not work at all in another.

In order to be really useful, this book should be adapted by you, the people who understand the problems of water supply and sanitation in the communities where you live.

Please feel free to copy, reproduce or adapt parts or the whole of the manual. We would like it to be changed to reflect or respond to different local needs. Any organisation or individual who wishes to copy, reproduce or adapt this manual for profit/commercial purposes must first apply to TALC, the publisher.

Anyone who wishes to use any part of this manual for preparing training materials is welcome to do so. Please remember to acknowledge this manual as your source. There is no need to request permission from either the authors or the publisher, as long as the sections that are reproduced are distributed free or at low cost.

For this reason, we are requesting the help of you, the user. Your experience, different approaches, practice or suggestions could contribute to a better manual – and one that would better serve the water supply and sanitation needs of more communities.

Please tell us:

- What you found useful
- What was not useful, or did not work, from your experience
- What could be added to make the manual more useful to others

Please send your examples, suggestions, criticisms or improvements either by post to us at:

Water and Sanitation, TALC, P.O. Box 49, St. Albans, Herts., AL1 5TX, U K

or by email to:
water.sanitation@talcuk.org

Thank you,

The authors and TALC

DAILY ACTIVITIES / TIMES OF DAY / STAGES OF HOUSE BUILDING

Poster G.1

DAWN/GETTING UP

Poster G.2
EATING A MEAL

DAILY ACTIVITIES

Poster G.5

CLEANING AROUND THE HOUSE

DAILY ACTIVITIES

DRINKING BEER

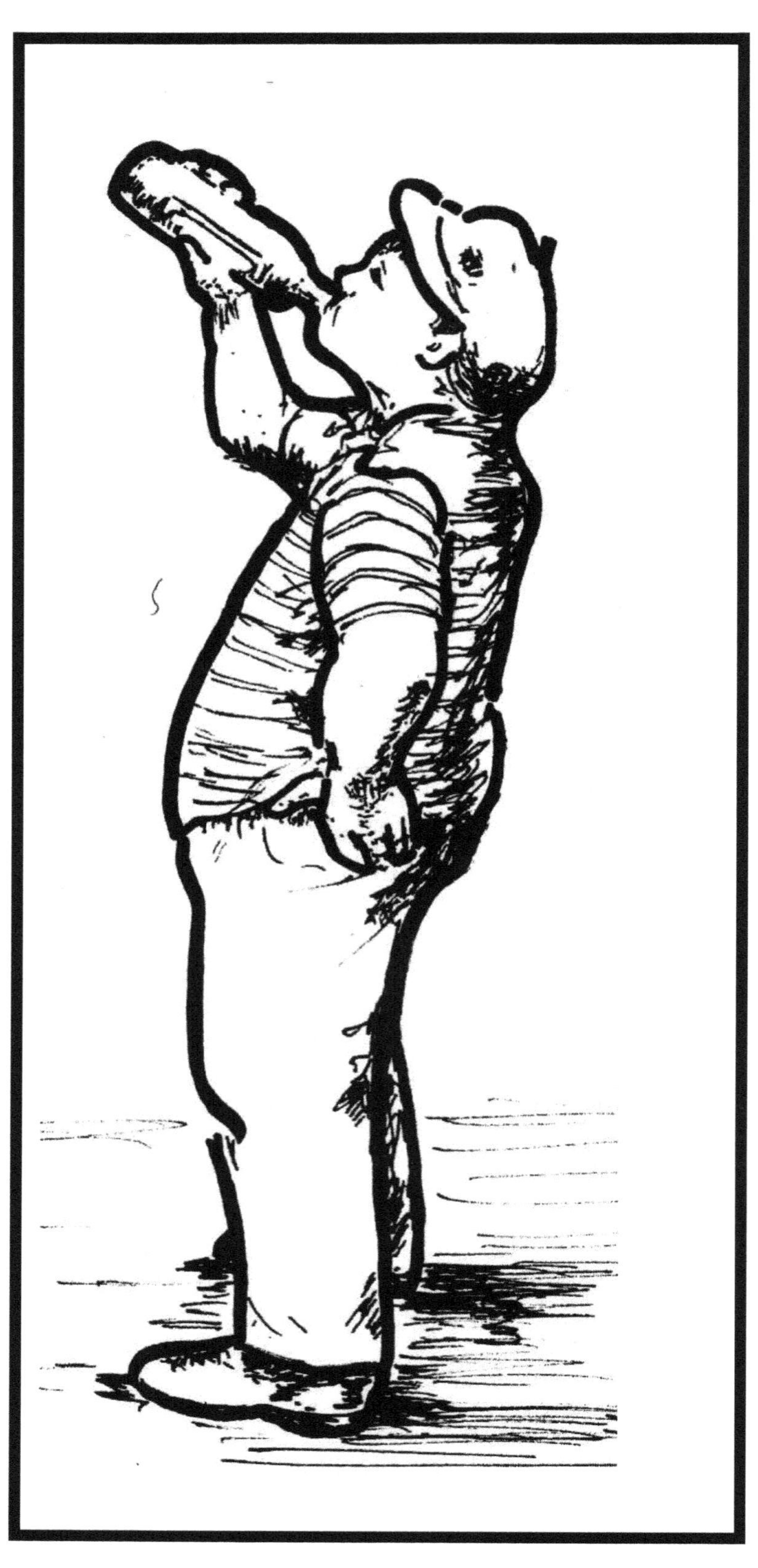

DAILY ACTIVITIES

DAILY ACTIVITIES

Poster G.8
WASHING CLOTHES

Poster G.11
BATHING

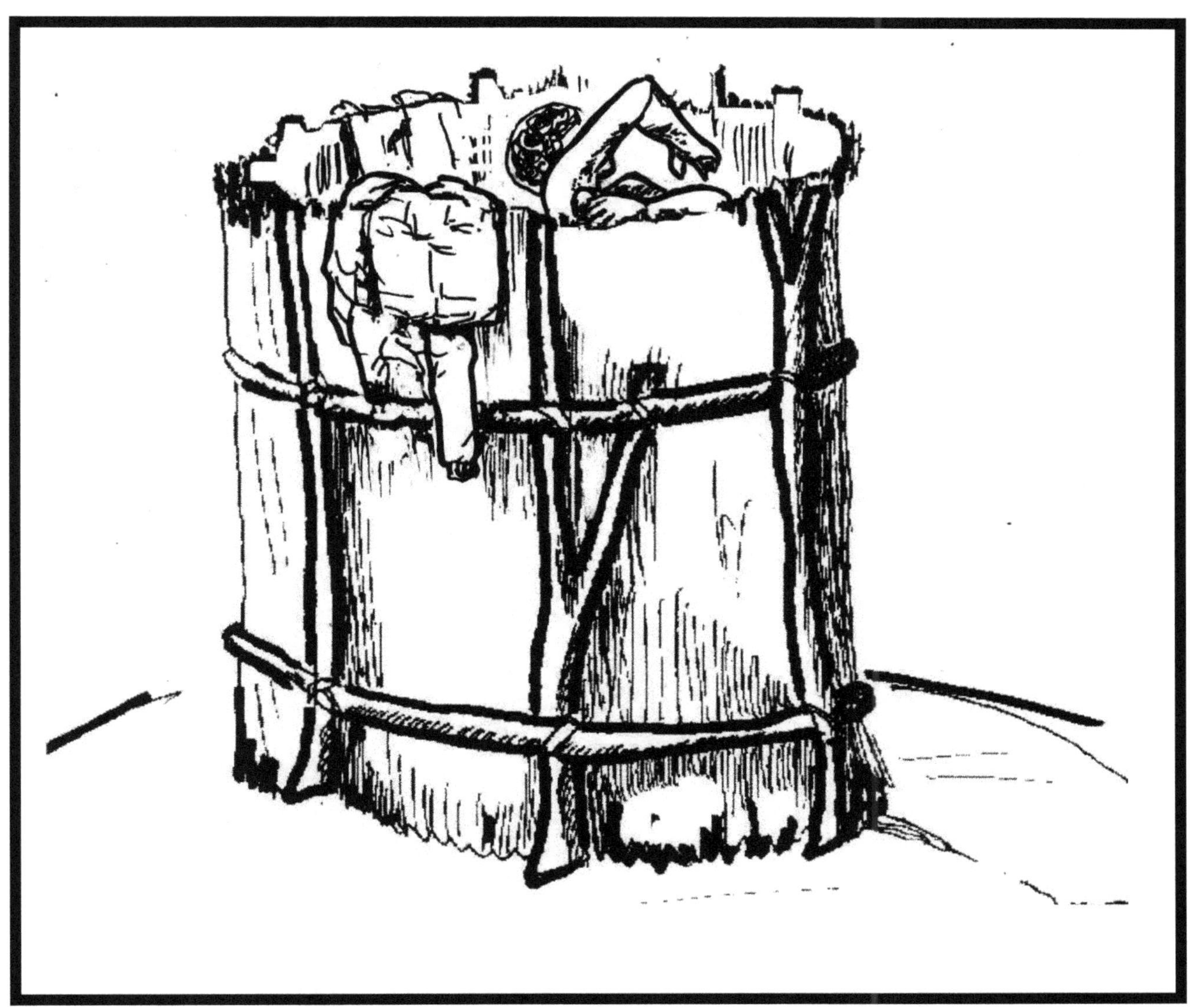

DAILY ACTIVITIES

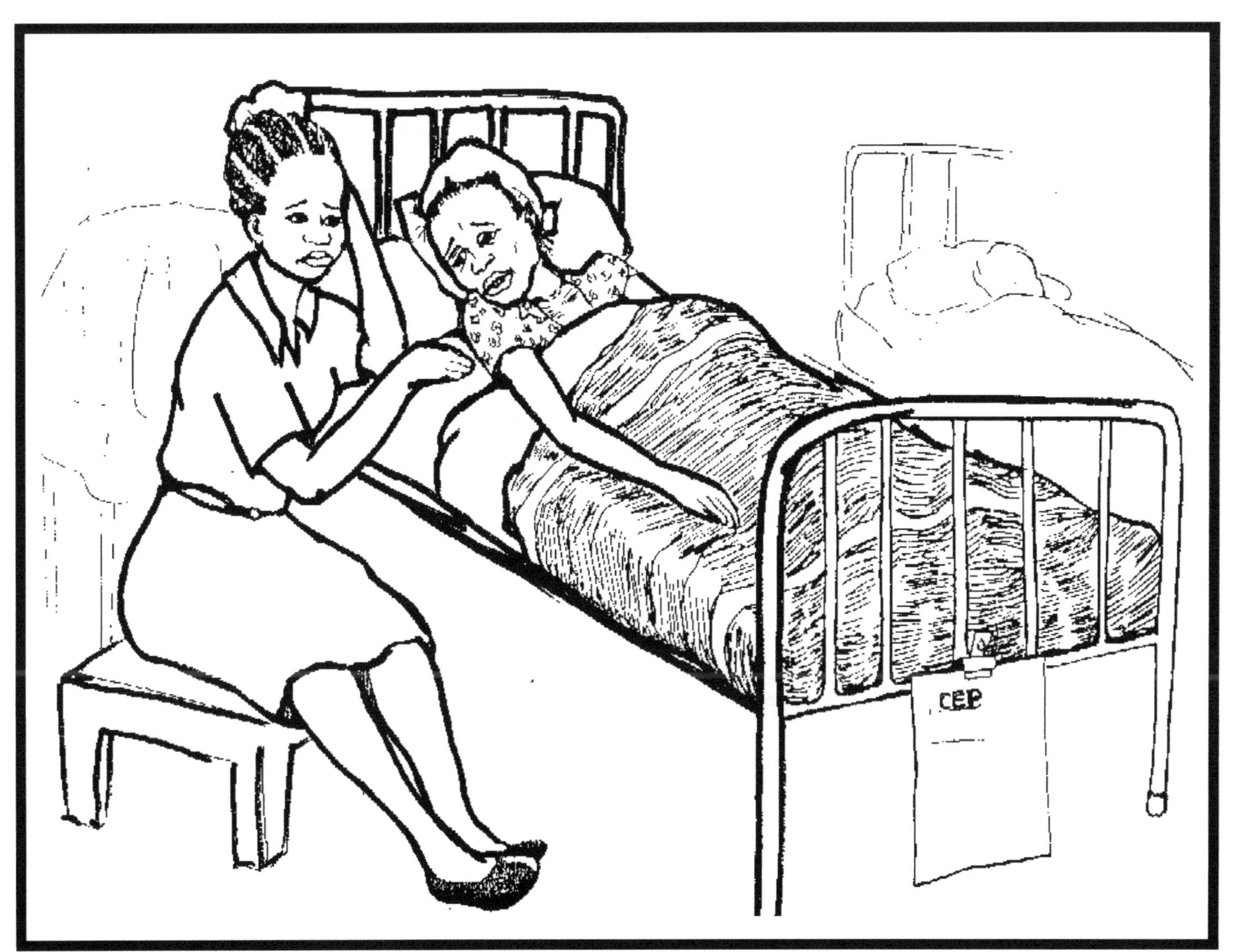
CEP

HYGIENE PRACTICES –
FOR FAECAL TRANSMISSION DIAGRAM AND 3-PILE SORTING

Poster H.1

HAND

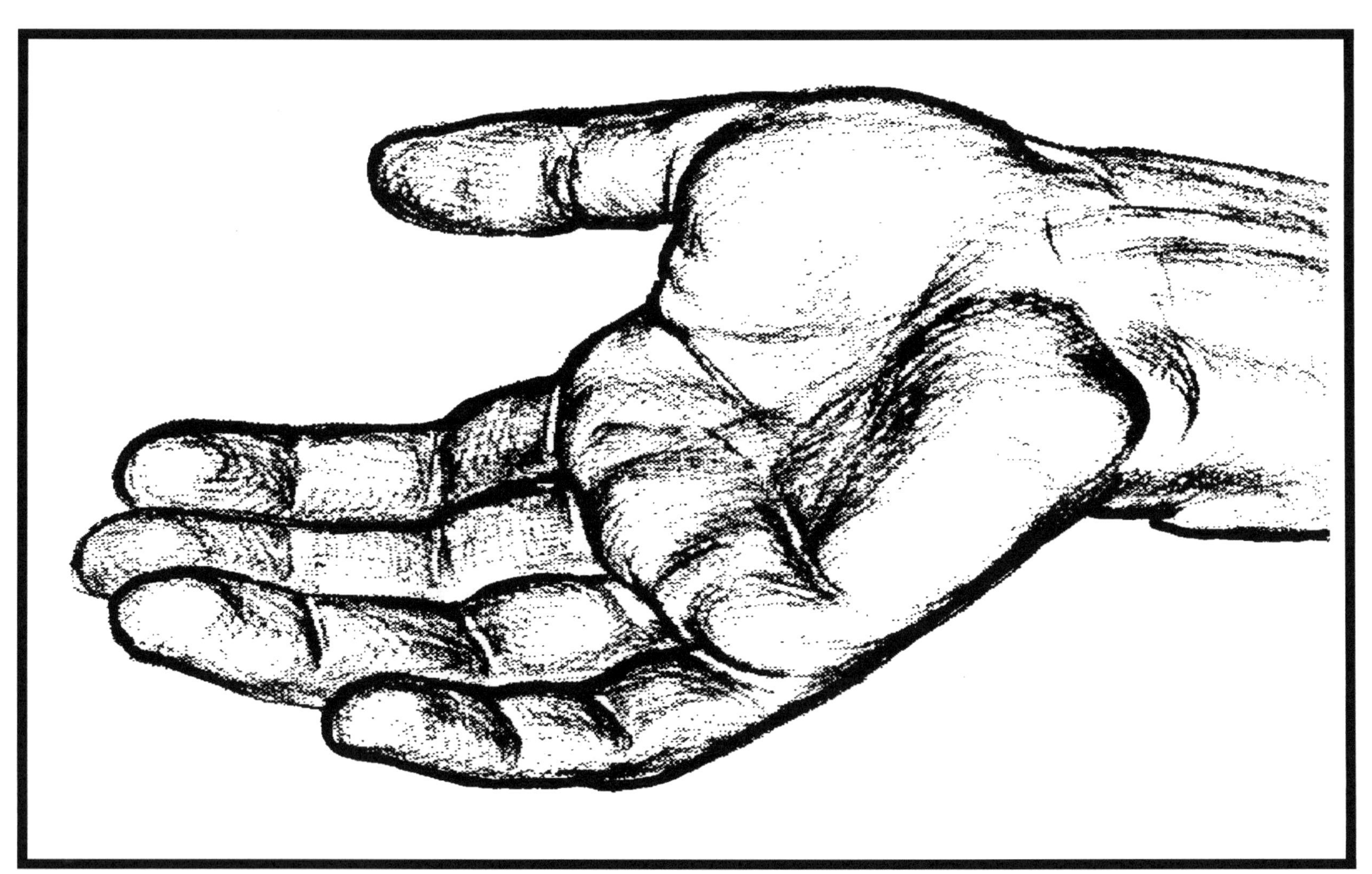

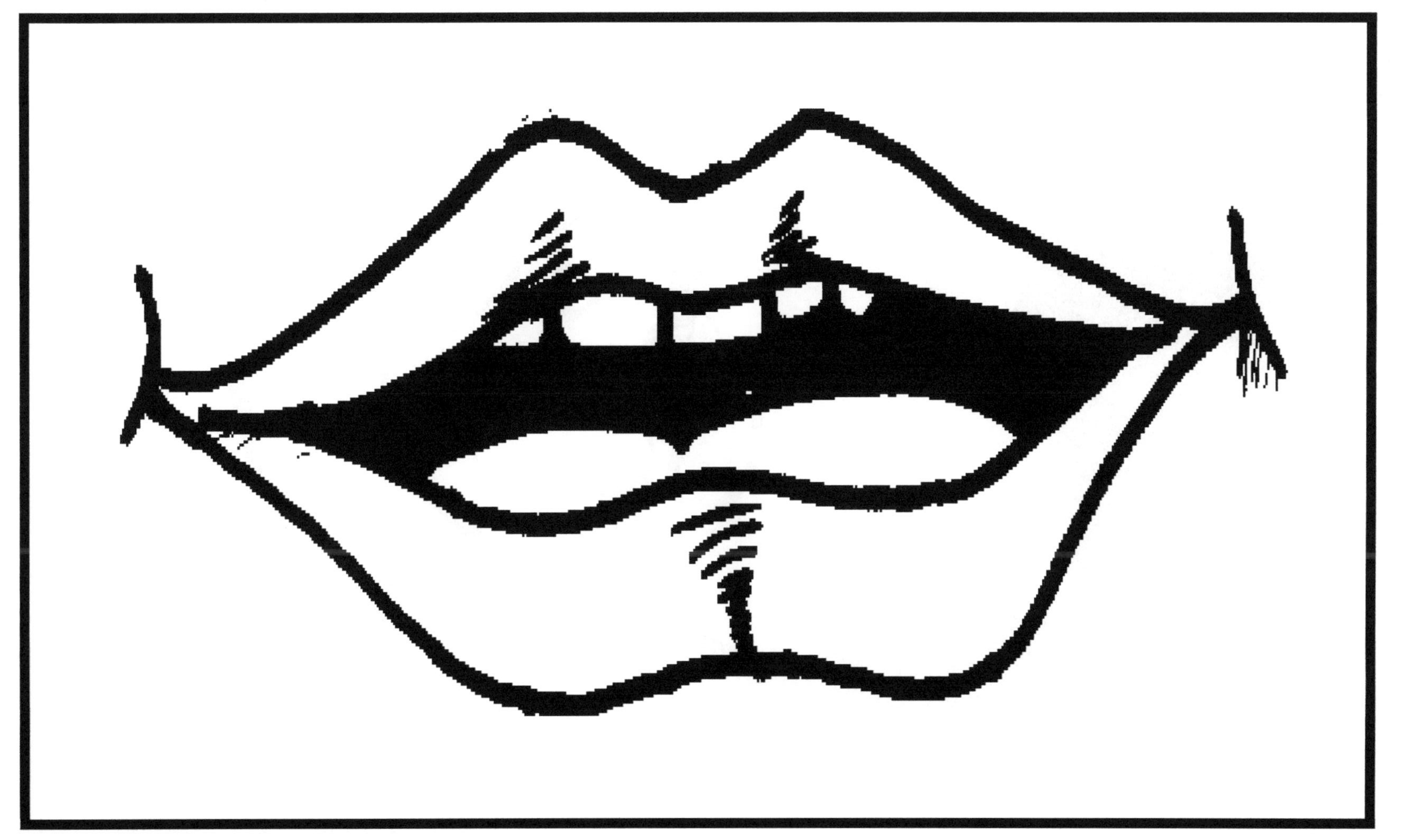

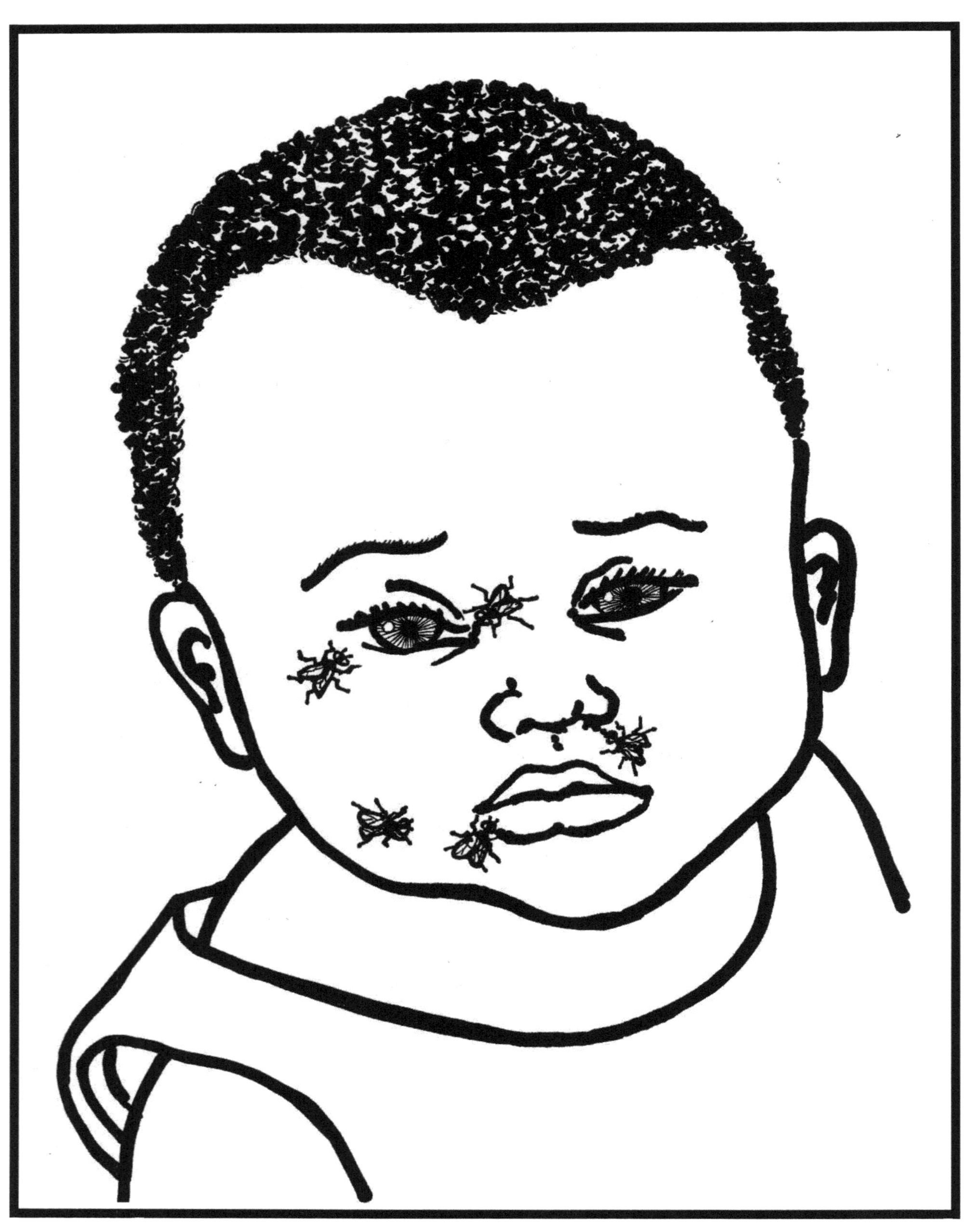

Poster H.4

DOG LICKING THE PLATES

Poster H.5

UNCOVERED FOOD WITH FLIES

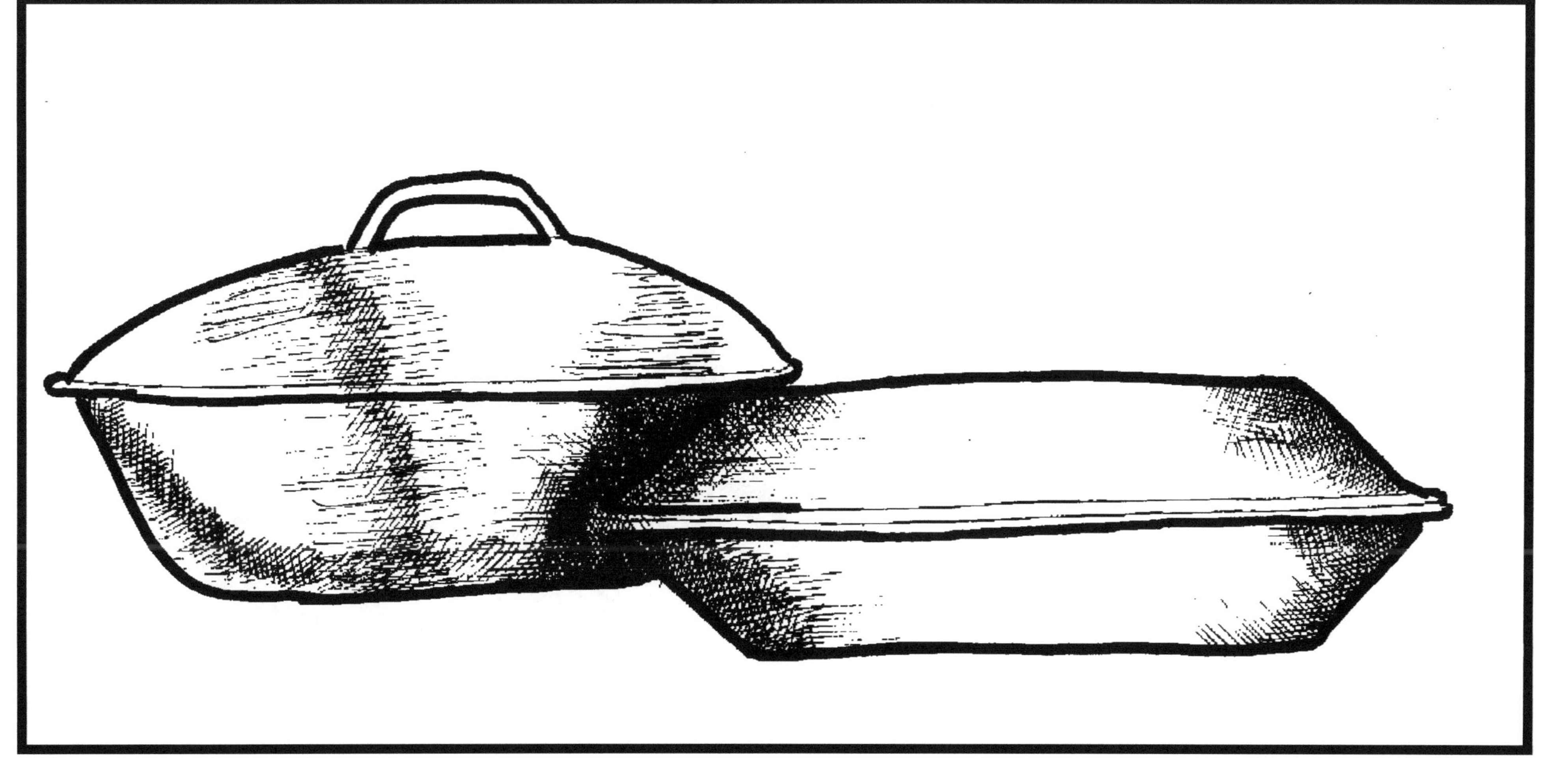

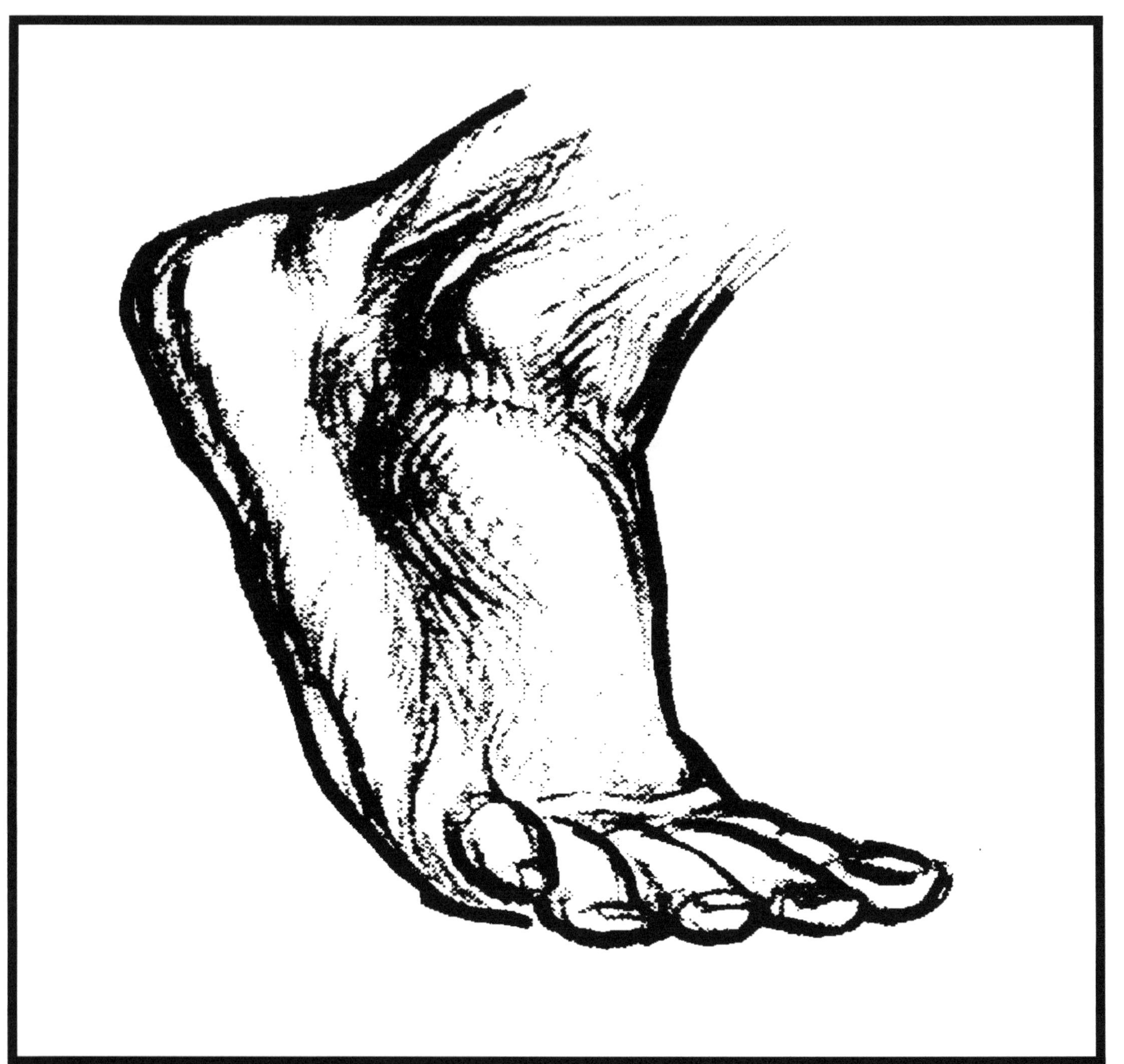

HYGIENE PRACTICES

Poster H.9

THROWING RUBBISH INTO A PIT

Poster H.11

DIPPING WATER CONTAINER INTO A WELL

Poster H.12

WASHING CHILDREN'S HANDS

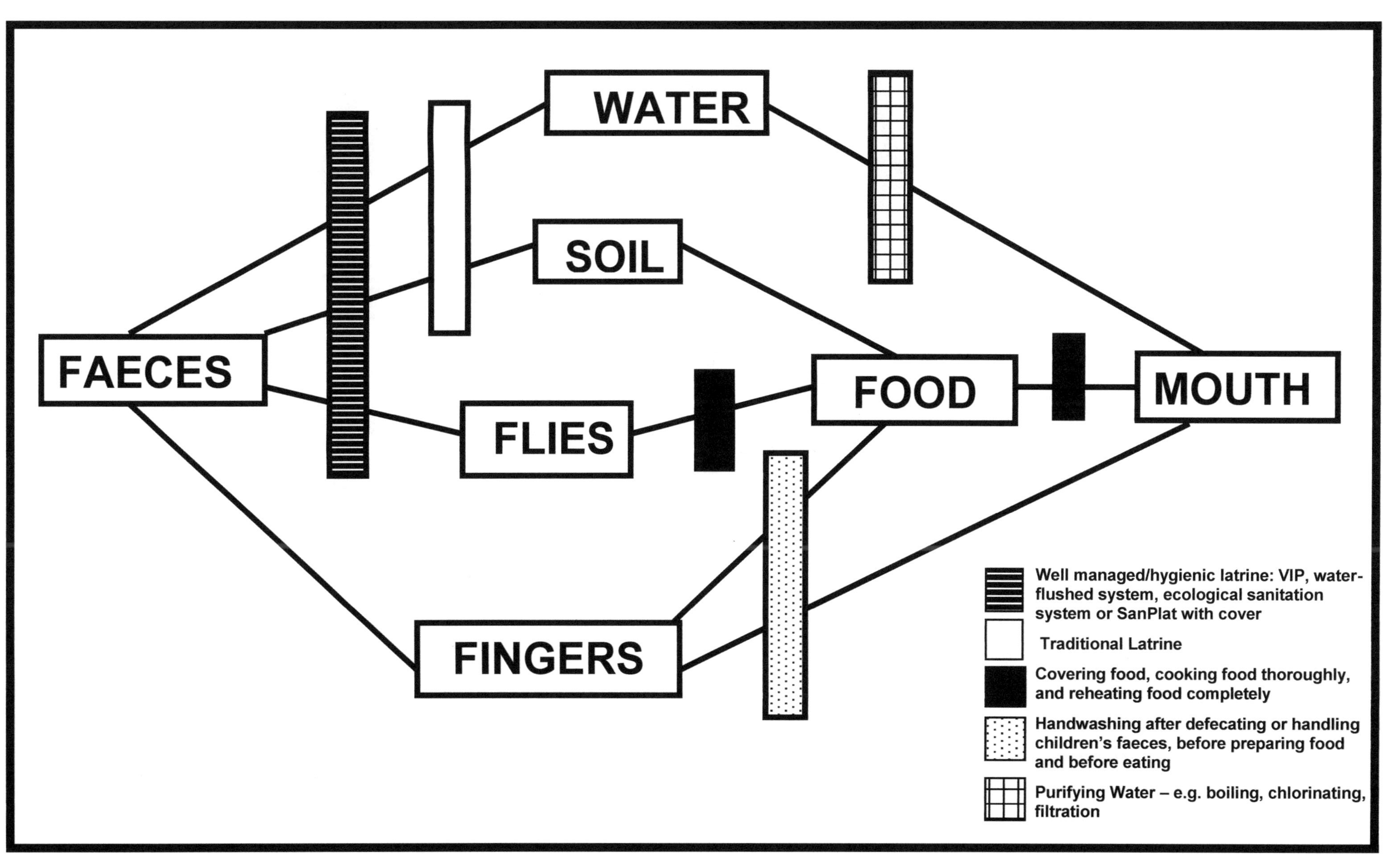
WATER
SOIL
FAECES
FLIES
FOOD
MOUTH
FINGERS
Well managed/hygienic latrine: VIP, water-flushed system, ecological sanitation system or SanPlat with cover
Traditional Latrine
Covering food, cooking food thoroughly, and reheating food completely
Handwashing after defecating or handling children's faeces, before preparing food and before eating
Purifying Water – e.g. boiling, chlorinating, filtration

HANDWASHING PRACTICES

Poster H.21
INTO BOWL, WITHOUT SOAP

HANDWASHING PRACTICES

INTO BOWL, WITH SOAP

Poster H.23
SCOOPING OUT OF BOWL,
NO SOAP

Poster H.24
SCOOPING OUT OF BOWL, USING SOAP

HANDWASHING PRACTICES

I feel proud when my children are admired.
Washing their hands with soap makes them smell nice
and clean and adorable.

Using a bottle with a handle

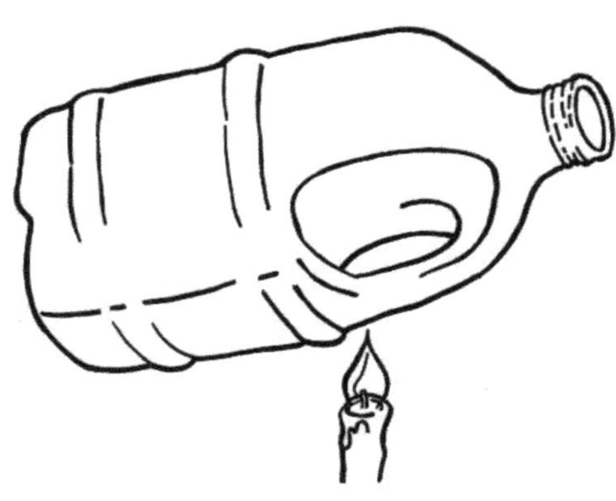

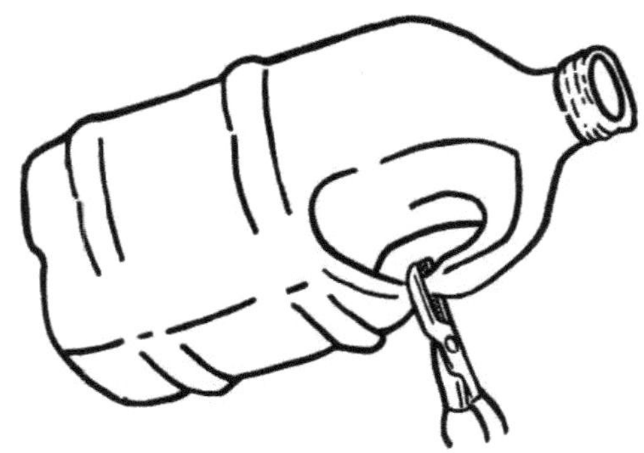

1. Warm the base of the bottle handle over a candle until soft.

2. Pinch the soft part of the handle with pliers to seal it and prevent water flowing through.

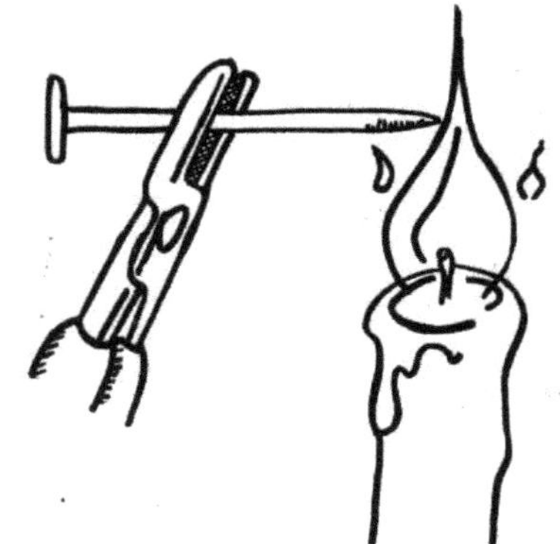

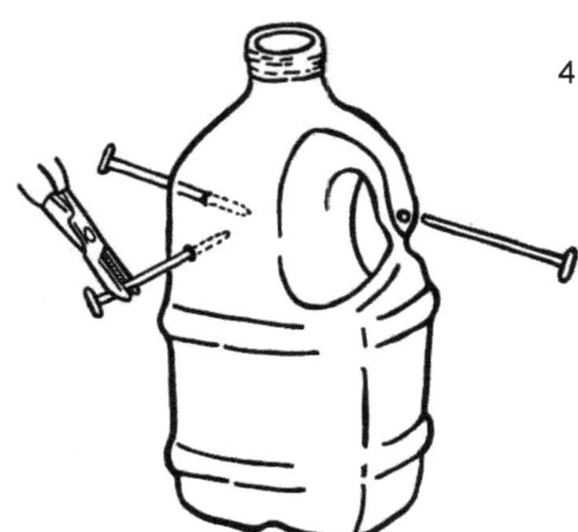

3. Heat a small nail over a candle using pliers.

4. Use the hot nail to make a small hole on the outside edge of the handle, just above the sealed area. Reheat it and make two larger holes in the back of the bottle about a thumb-width apart.

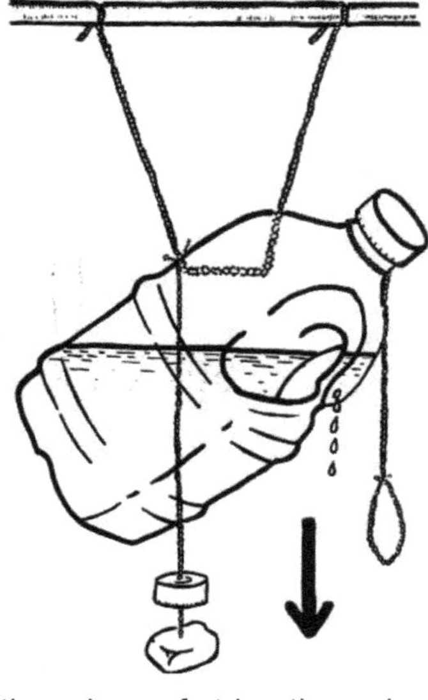

5. Thread a string through the two back holes and tie to a stick Pour water into the tippy tap until it is almost level with the holes in the back of the bottle.

6. Thread another piece of string through a bar of soap and an empty tin can (the base facing upwards to protect the soap from rain and sun. Attach to the tippy-tap string as shown. Tie a third string to the bottle cap and leave end hanging – pulling this string tips the bottle and causes water to flow from the hole in the handle.

Using a bottle without a handle

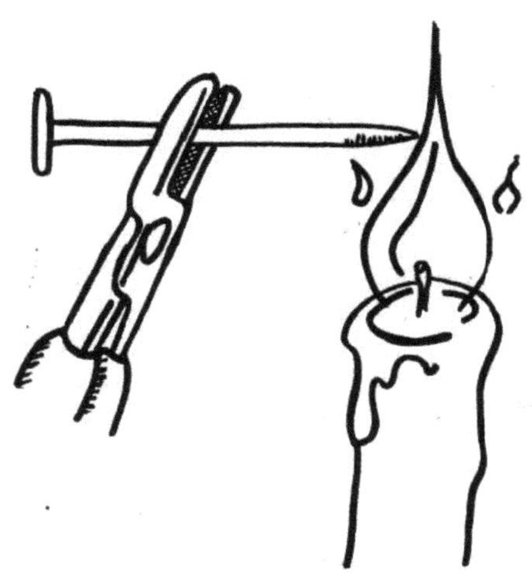

1. Heat the point of a small nail over a candle (make sure that you use pliers to hold the nail).

2. Use the hot nail to make two small holes on each side at the top of the bottle. One hole is for the water to get out and one allows air to get in.

3. Take a long piece of string and tie one end around the neck of the bottle. Tie the other around the body of the bottle, near the base.

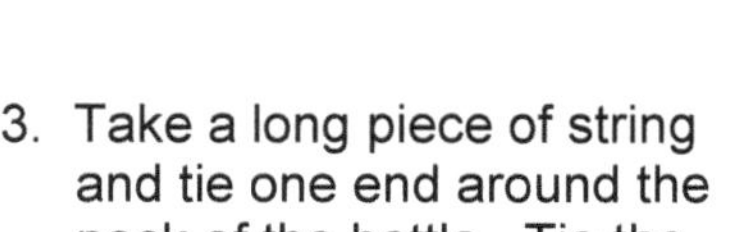

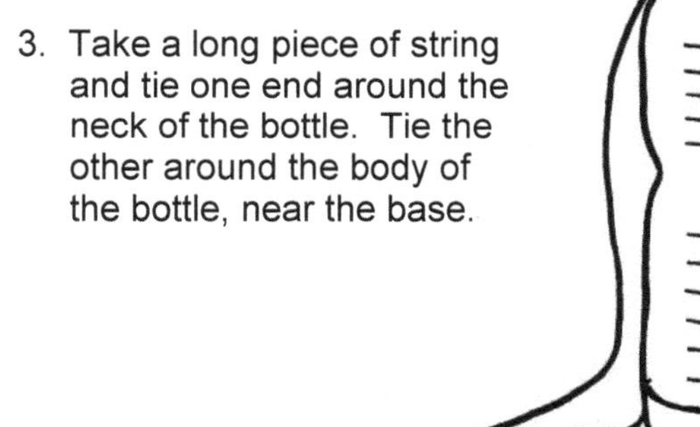

4. Pour water into the bottle. Stop before the water level reaches the holes.

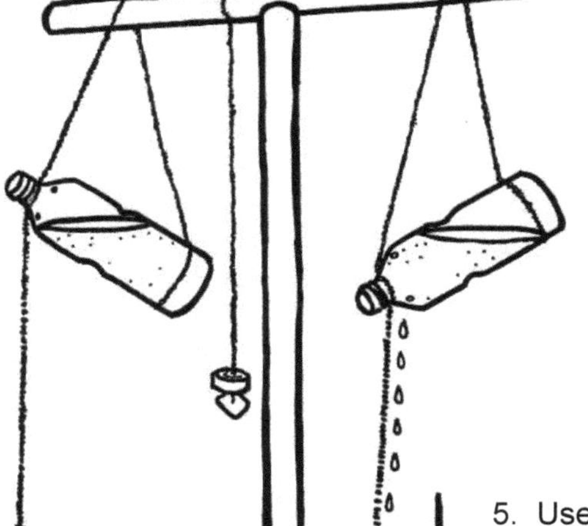

5. Use the string to hang the bottle in front of the latrine door, or where it will best remind the toilet users to wash their hands before leaving the toilet. Tilting the bottle up or down is like opening and closing a tap. Hang a piece of soap next to the bottle.

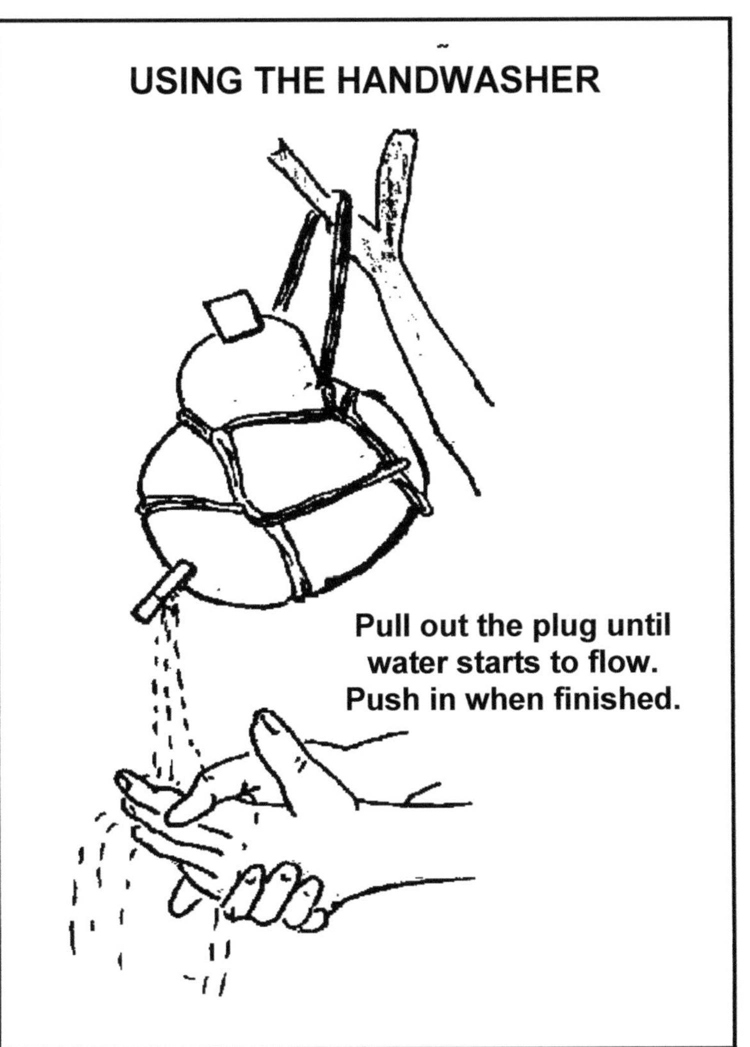

FILLING THE CALABASH HANDWASHER
Small hole with
plug in place
Wood plug with groove
(Shown large)
USING THE HANDWASHER
Pull out the plug until
water starts to flow.
Push in when finished.

LEAKING LADLE HANDWASHER

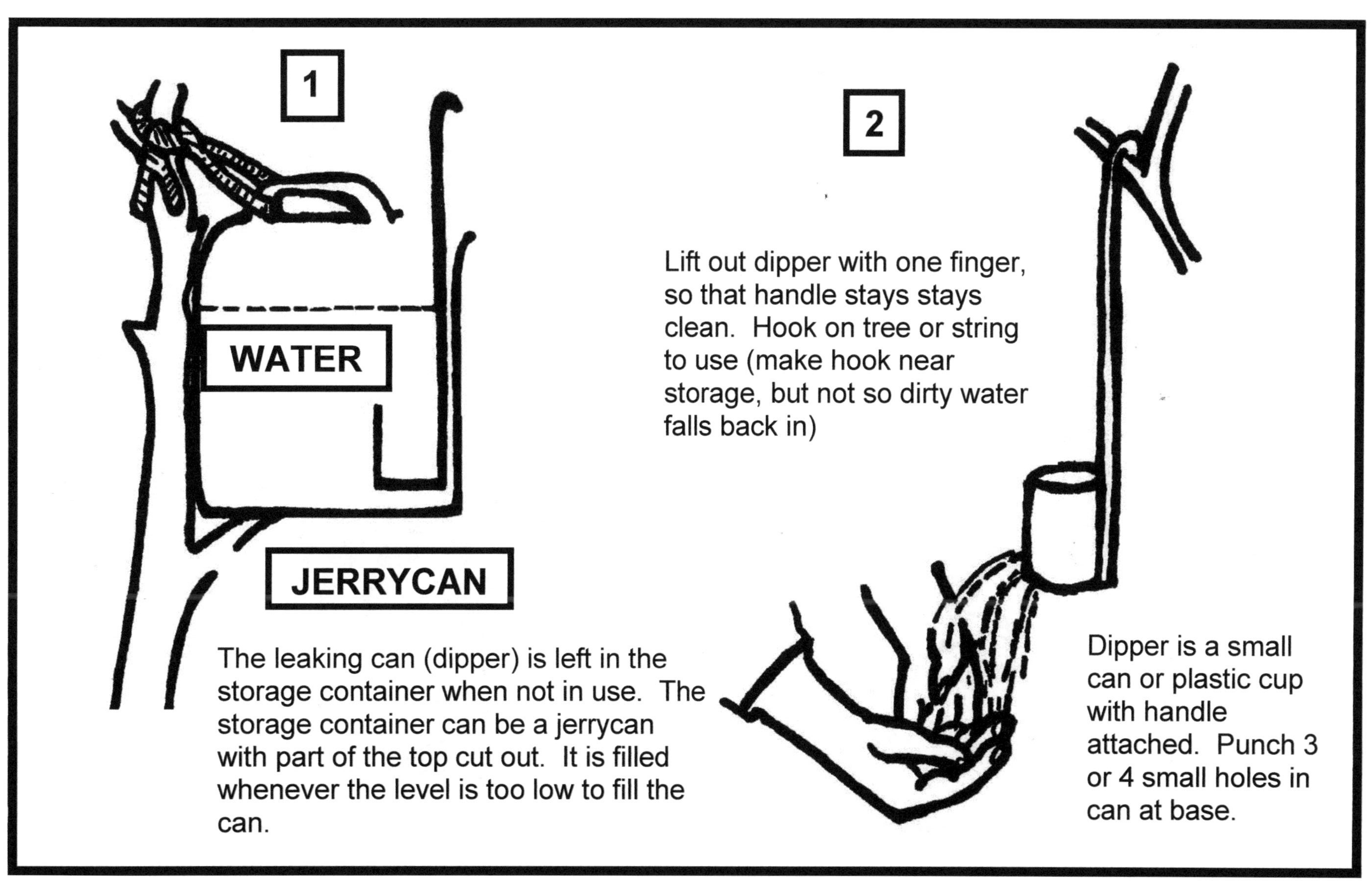

The leaking can (dipper) is left in the storage container when not in use. The storage container can be a jerrycan with part of the top cut out. It is filled whenever the level is too low to fill the can.

Lift out dipper with one finger, so that handle stays stays clean. Hook on tree or string to use (make hook near storage, but not so dirty water falls back in)

Dipper is a small can or plastic cup with handle attached. Punch 3 or 4 small holes in can at base.

GOOD HAND HYGIENE

1. Rub palms together

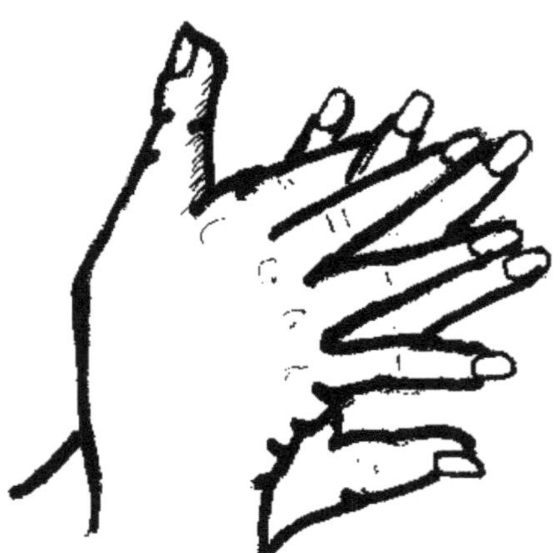

2. Right palm against back of left hand. Left palm against back of right hand.

3. Palm against palm with fingers interlocked.

4. On the inside of the fingers, with palms opposed and with fingers interlocked.

Poster H.34

CONSTRUCTIONS FOR HANDWASHING

TYPES OF LATRINE

Poster S.1

TRADITIONAL LATRINE (BASIC - UNIMPROVED)

TYPES OF LATRINE

TRADITIONAL LATRINE, IMPROVED (GRASS-ROOFED)

TYPES OF LATRINE

'TRADITIONAL' LATRINE IMPROVED (VIP)

TYPES OF LATRINE

TRADITIONAL, IMPROVED LATRINE

TYPES OF LATRINE

TYPES OF LATRINE

BRICK/MUD-BRICK IMPROVED PIT LATRINE, VIP

TYPES OF LATRINE

Poster S.8
IMPROVED PIT LATRINE (3) WITH SQUARE SANPLAT

TYPES OF LATRINE

VENTILATED IMPROVED PIT LATRINE (VIP) – How it works

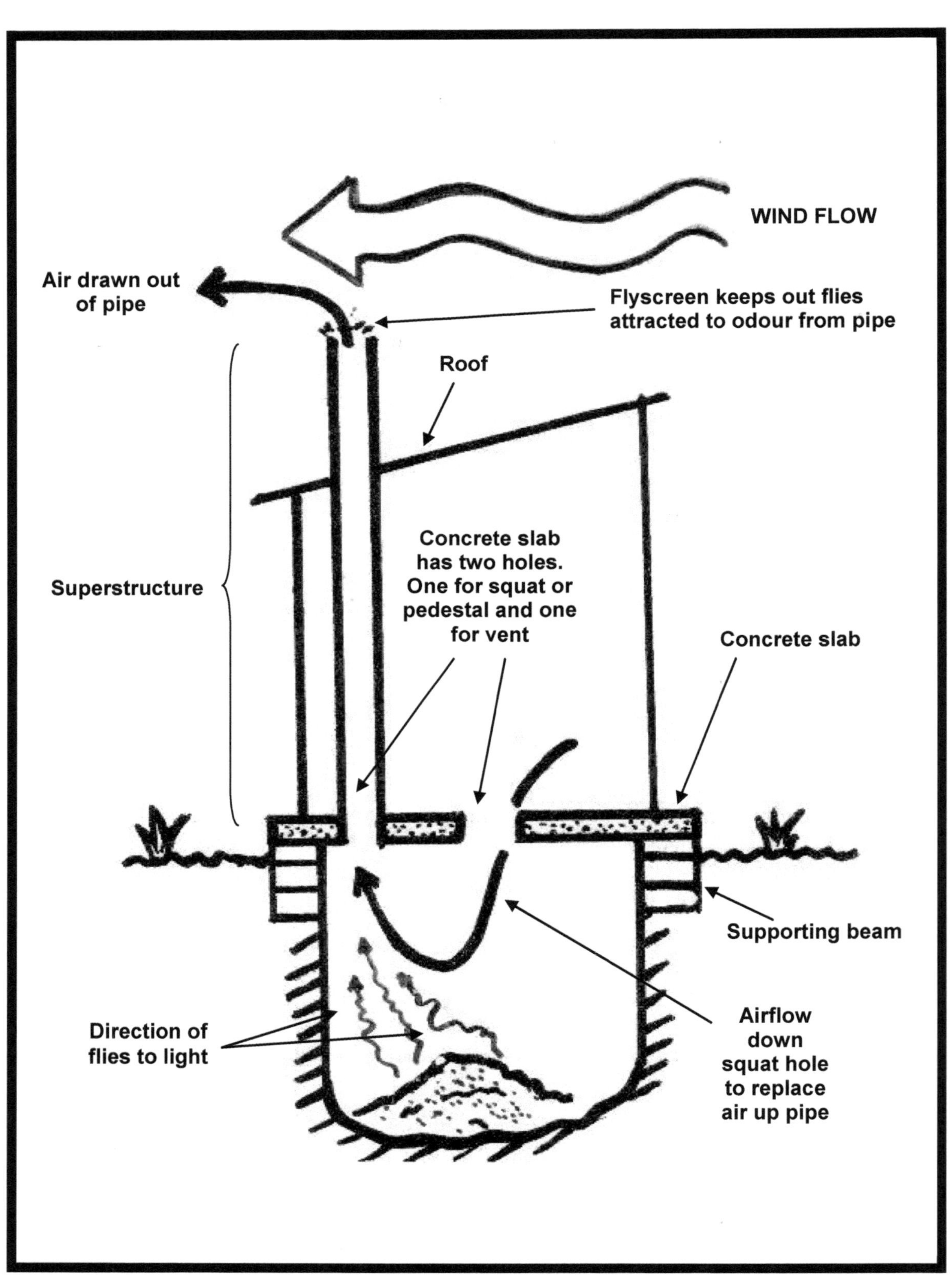

Poster S.10
ECOLOGICAL SANITATION/COMPOSTING/ URINE-DIVERTING LATRINE

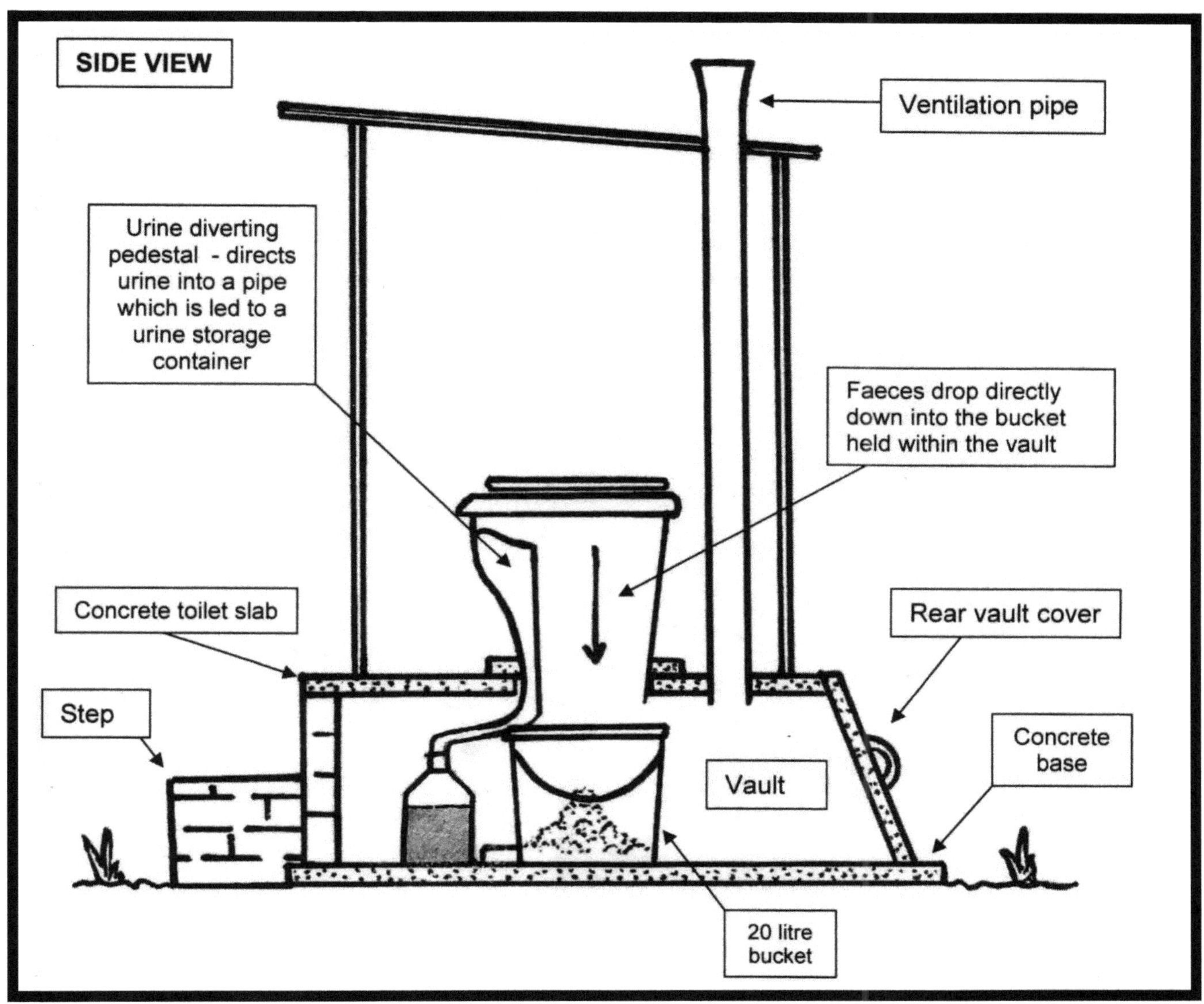

Urine and faeces are separated in a special pedestal or squat plate. Urine passes through a plastic pipe to a container on the side. Faeces fall into a bucket or vault, with wood ash and dry soil being added on top after every visit. When the bucket is nearly full, the contents are deposited in a composting site away from the toilet. The empty bucket is returned to the vault. Most urine diverting toilets are built above ground and therefore are useful if the ground water is high or if the ground is rocky

DEFECATION PRACTICES

DEFECATING IN A LAKE, DAM OR RIVER

DEFECATION PRACTICES

DEFECATION PRACTICES

Poster S.14
OPEN DEFECATION NEAR HOUSES

DEFECATION PRACTICES

Poster S.17
MOTHER CLEANS UP AFTER CHILD (CAT METHOD 2)

TYPES OF SQUAT HOLE

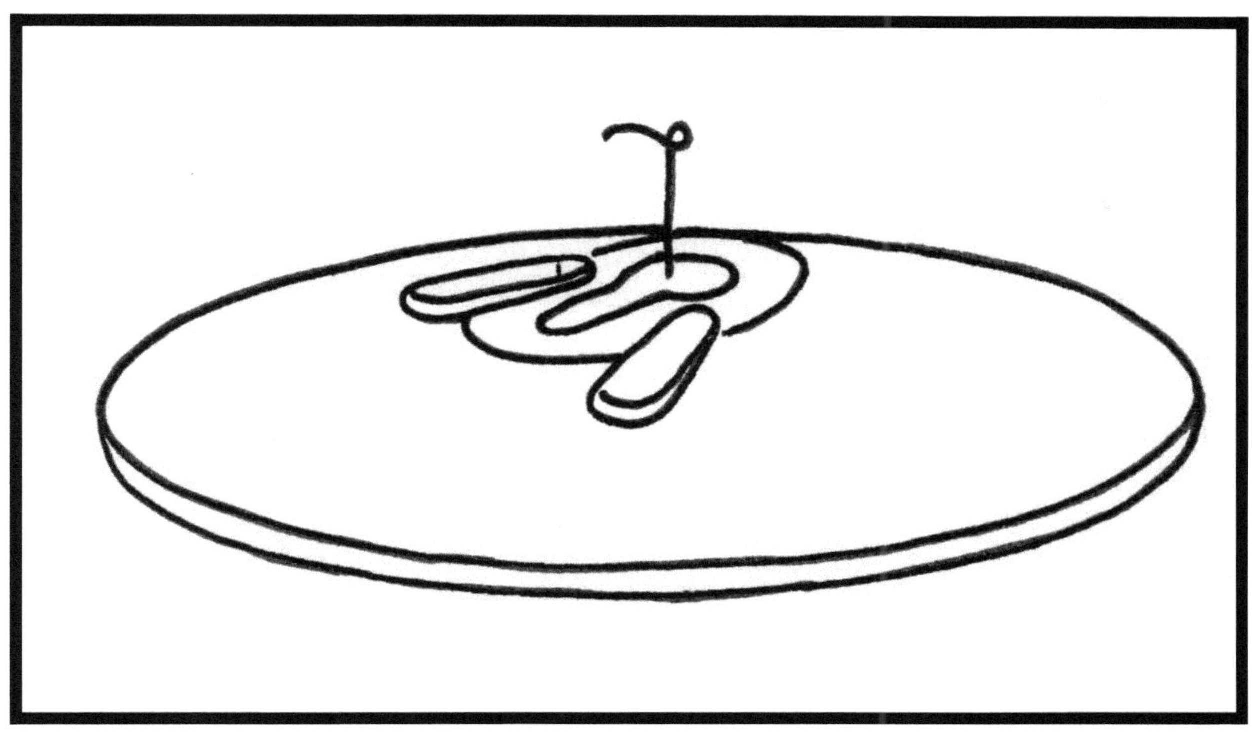

TYPES OF SQUAT HOLE

Poster S.21
SQUAT HOLE LINED WITH MORTAR OR CEMENT
CLAY/EARTH SLAB OVER WOOD

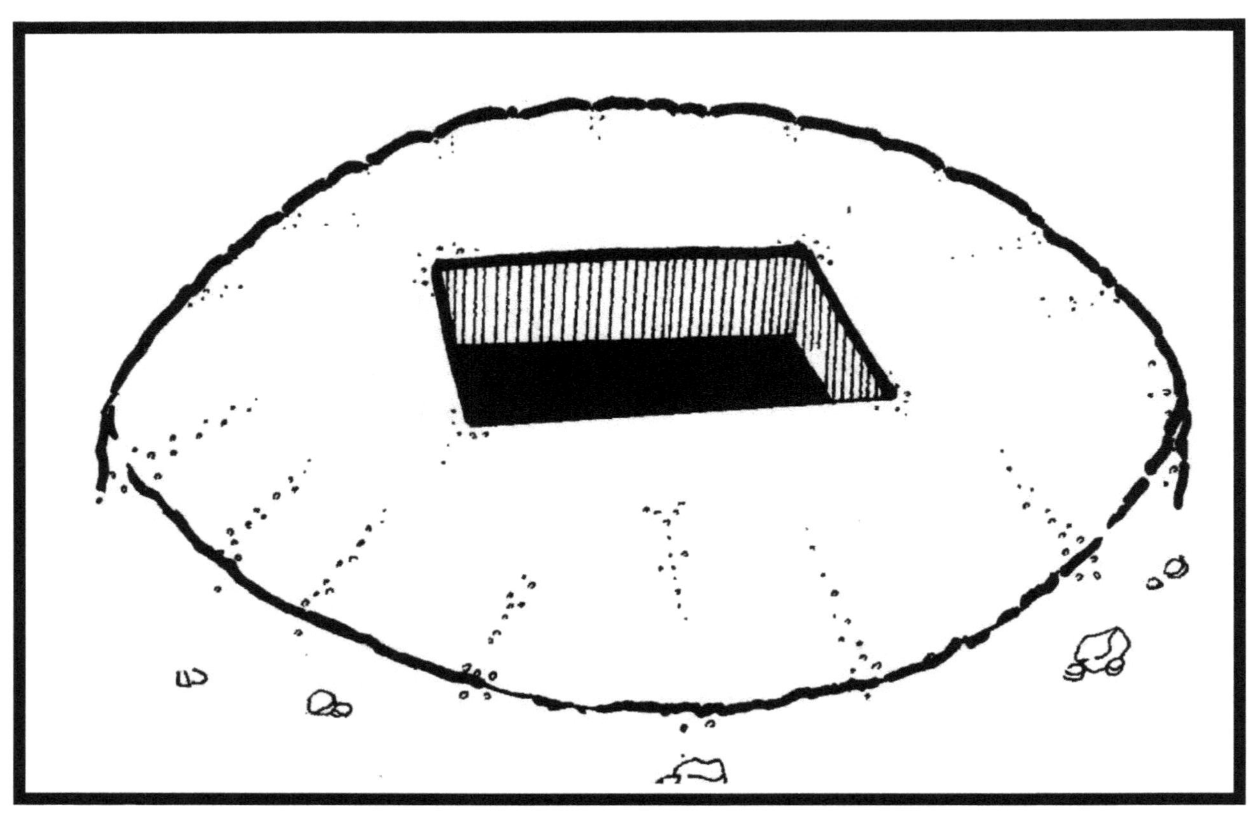

TYPES OF SQUAT HOLE

Poster S.22
WOODEN SLAB OVER SQUAT HOLE

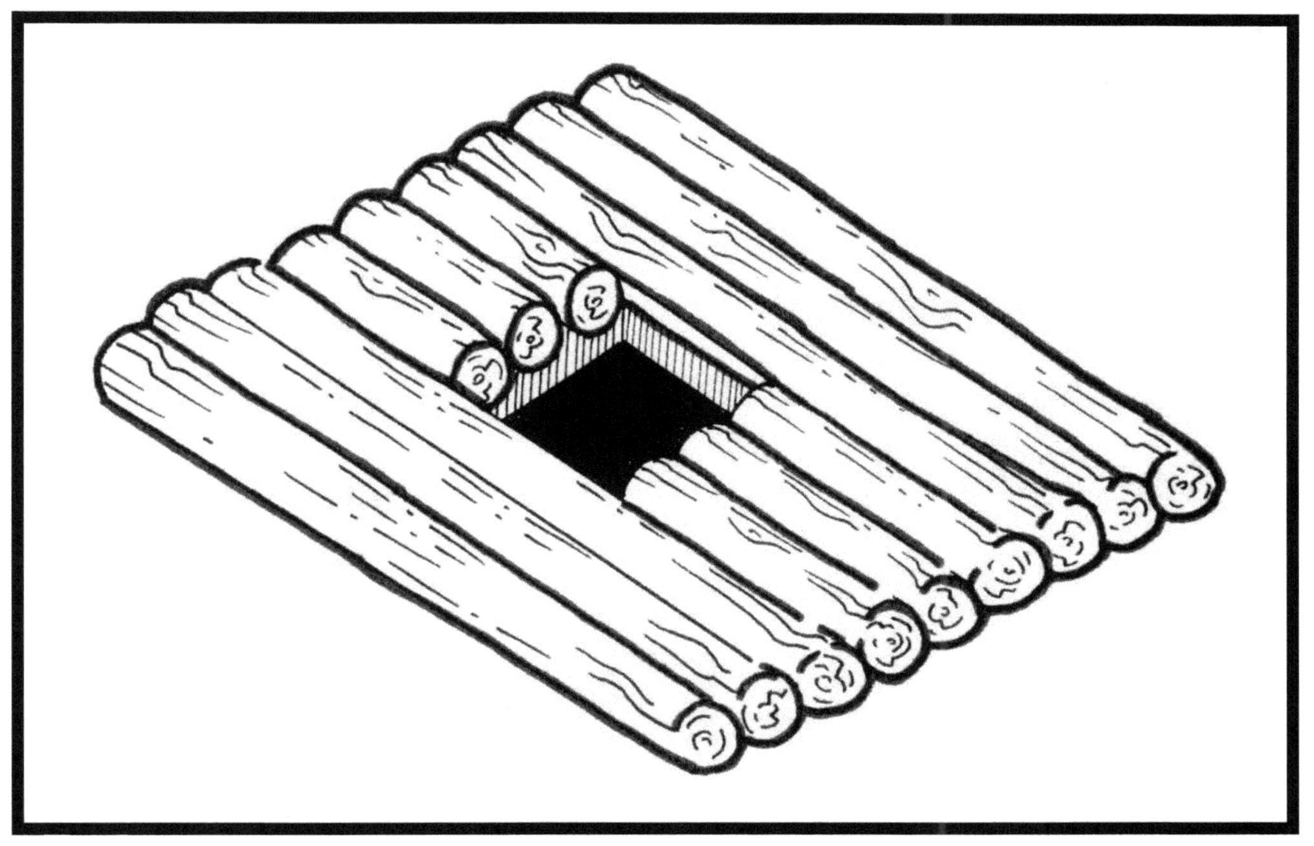

WATER SUPPLY:
MAIN (MOSTLY PROTECTED) SOURCE TYPES, INCLUDING WATER-LIFTING DEVICES

Poster W.1
HANDPUMP ON BOREHOLE

Poster W.2
HANDPUMP ON SHALLOW WELL

Poster W.3

BUCKET PUMP ON BOREHOLE/ HAND-AUGERED WELL

Poster W.5
IMPROVED FAMILY WELL (1)

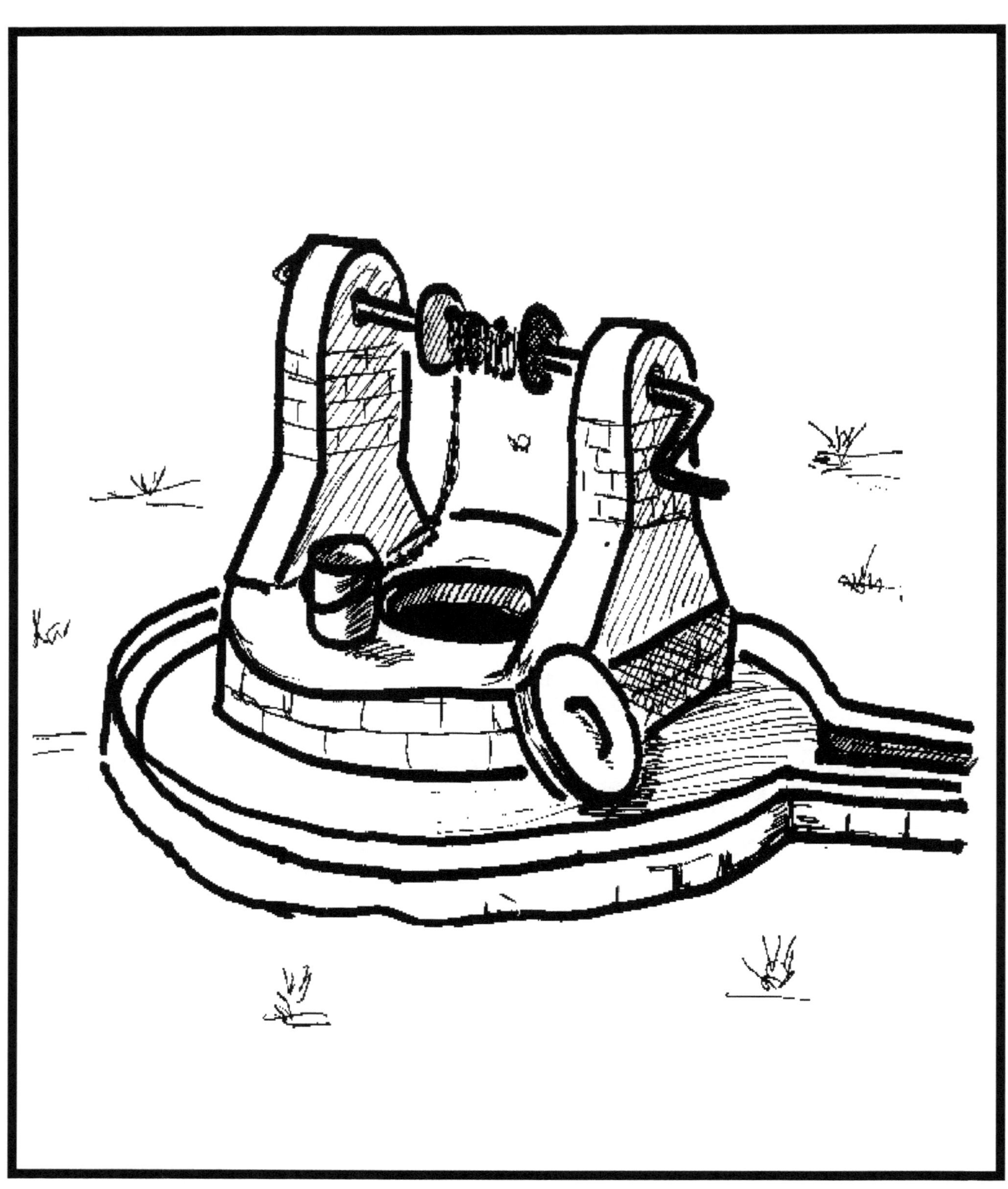

Poster W.6
IMPROVED FAMILY WELL (2)

WATER SUPPLY

WATER SUPPLY

Poster W.14
ROOF CATCHMENT OF RAINWATER

WATER SUPPLY

Poster W.15

RAINWATER CATCHMENT (especially Schools and Rural Health Centres)

BUCKET AND BEAM LIFTING DEVICE

WATER SUPPLY

Poster W.17

ROPE PUMP

Poster W.18
TREADLE PUMP

SOME WAYS OF TRANSPORTING WATER

Poster W.20

ON THE HEAD AND BY HAND (1)

WAYS TO TRANSPORT WATER

Poster W. 22

BY HAND

WAYS TO TRANSPORT WATER

Poster W.23

ALL THE FAMILY, CARRYING WATER ON HEAD AND BY HAND

WAYS TO TRANSPORT WATER

Poster W. 25

ON THE HEAD AND BY HAND (2)

WAYS TO TRANSPORT WATER

Poster W. 26
USING OXEN

WAYS TO TRANSPORT WATER

WAYS TO TRANSPORT WATER

OPTIONS FOR LINING SHAFTS (WELLS AND LATRINES)

Poster W. 29

SHALLOW BRICK LINING OR MASONRY FOR SPRING, SCOOPHOLE OR LATRINE PIT

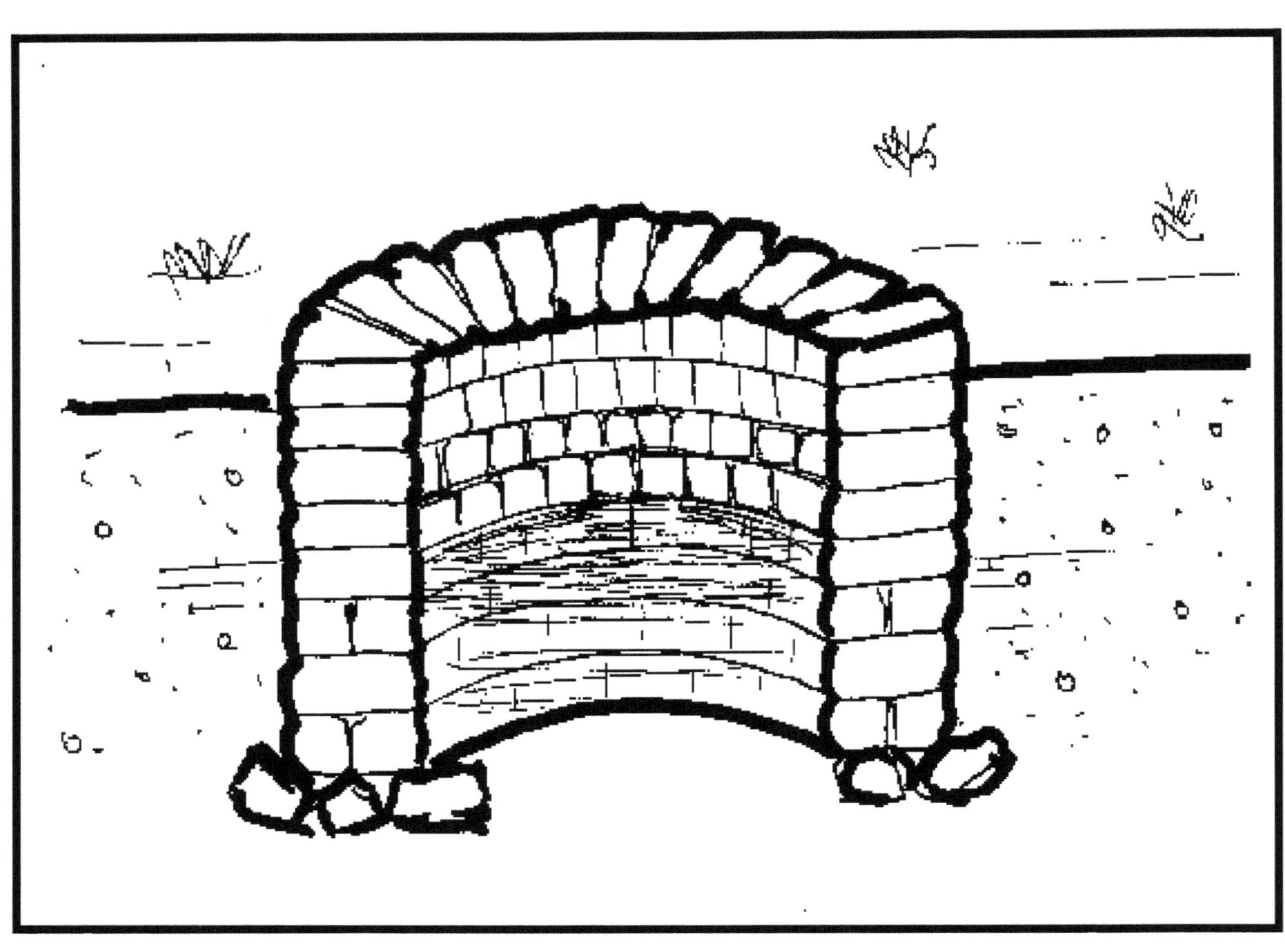

OPTIONS FOR LINING SHAFTS

Poster W. 30

FULLY BRICK-LINED WELL WITH CONCRETE RINGS BELOW THE HIGHEST WATER LEVEL

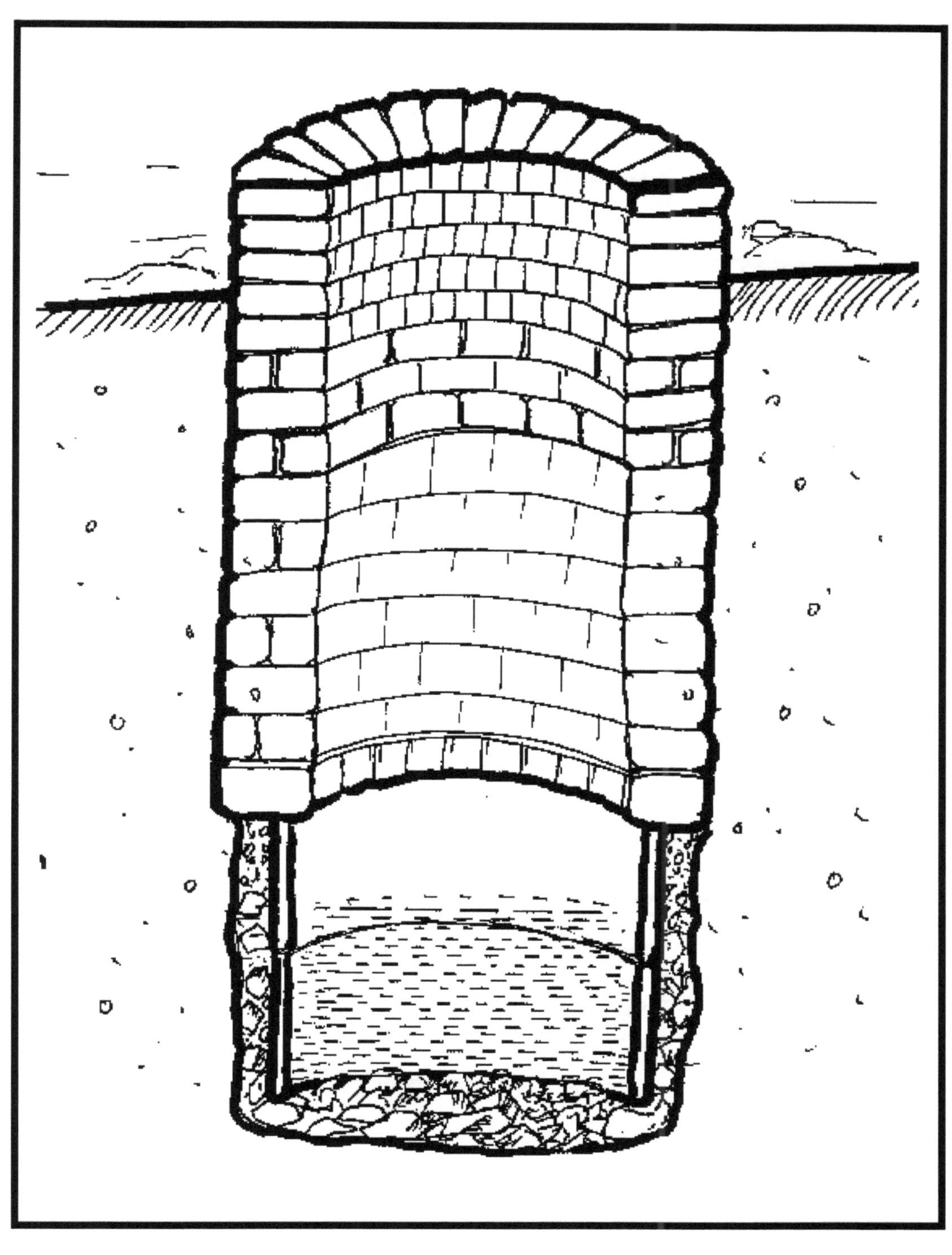

OPTIONS FOR LINING SHAFTS

FULLY LINED WELL OF BRICK AND STONE, WITH 'DRY STONE' BELOW THE HIGHEST WATER LEVEL

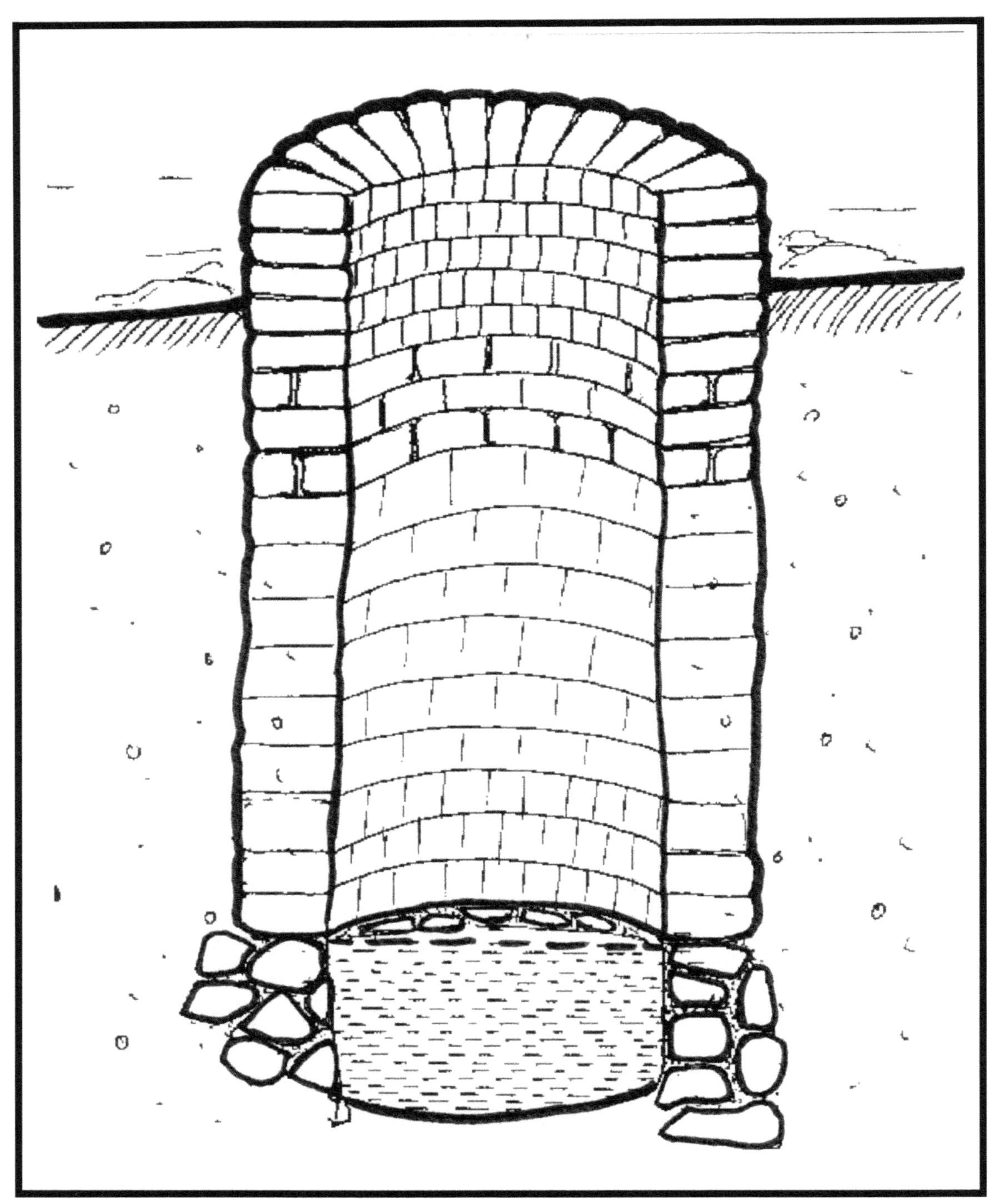

Bricks should not be used below water level unless they are extremely well-fired (and therefore strong) as water causes them to crumble. Preferable to use 'dry stone' (rocks or stones without mortar).

OPTIONS FOR LINING SHAFTS

PARTIAL BRICK LINING, OLD AND NEW

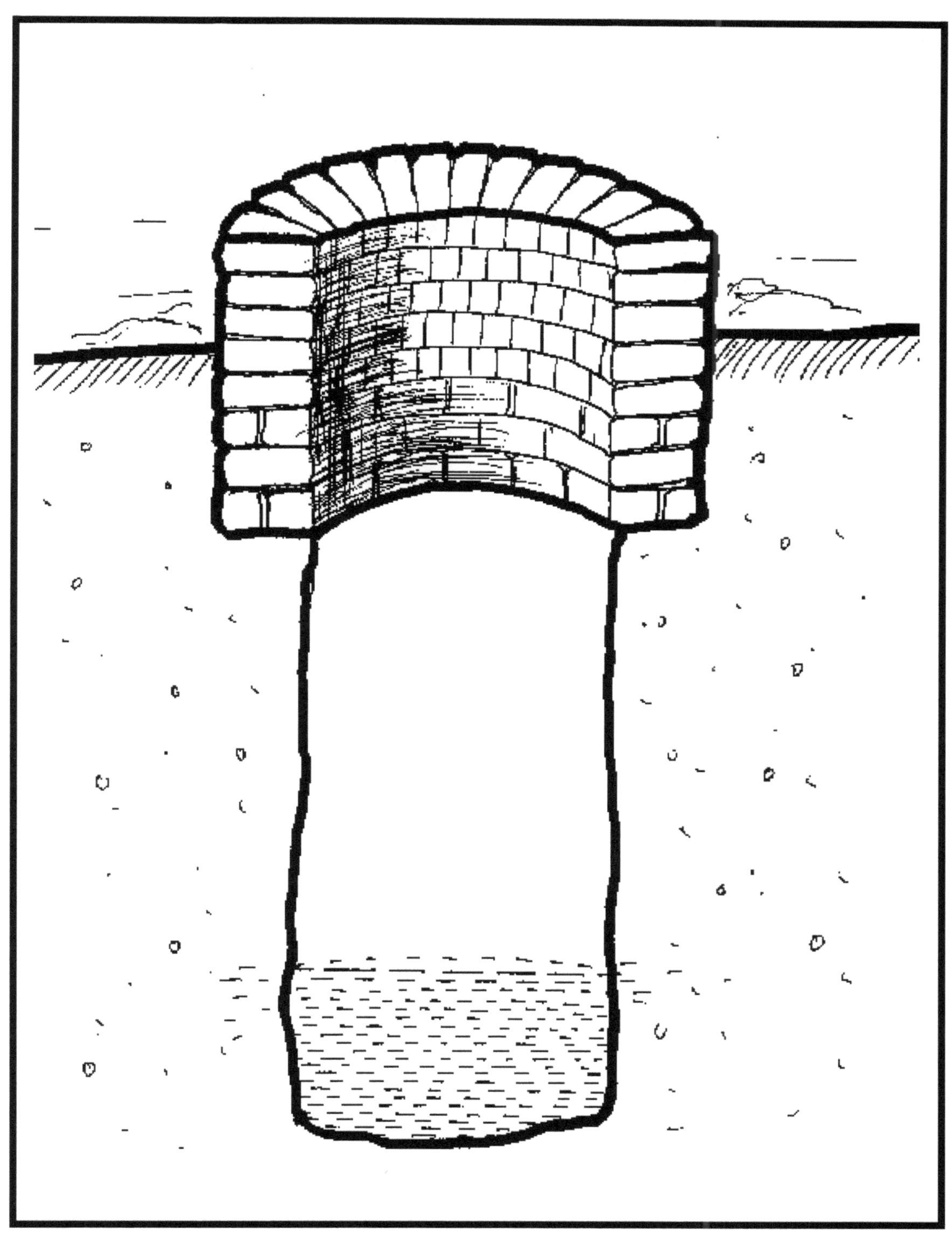

OPTIONS FOR LINING SHAFTS

PARTIAL CONCRETE RING LINING

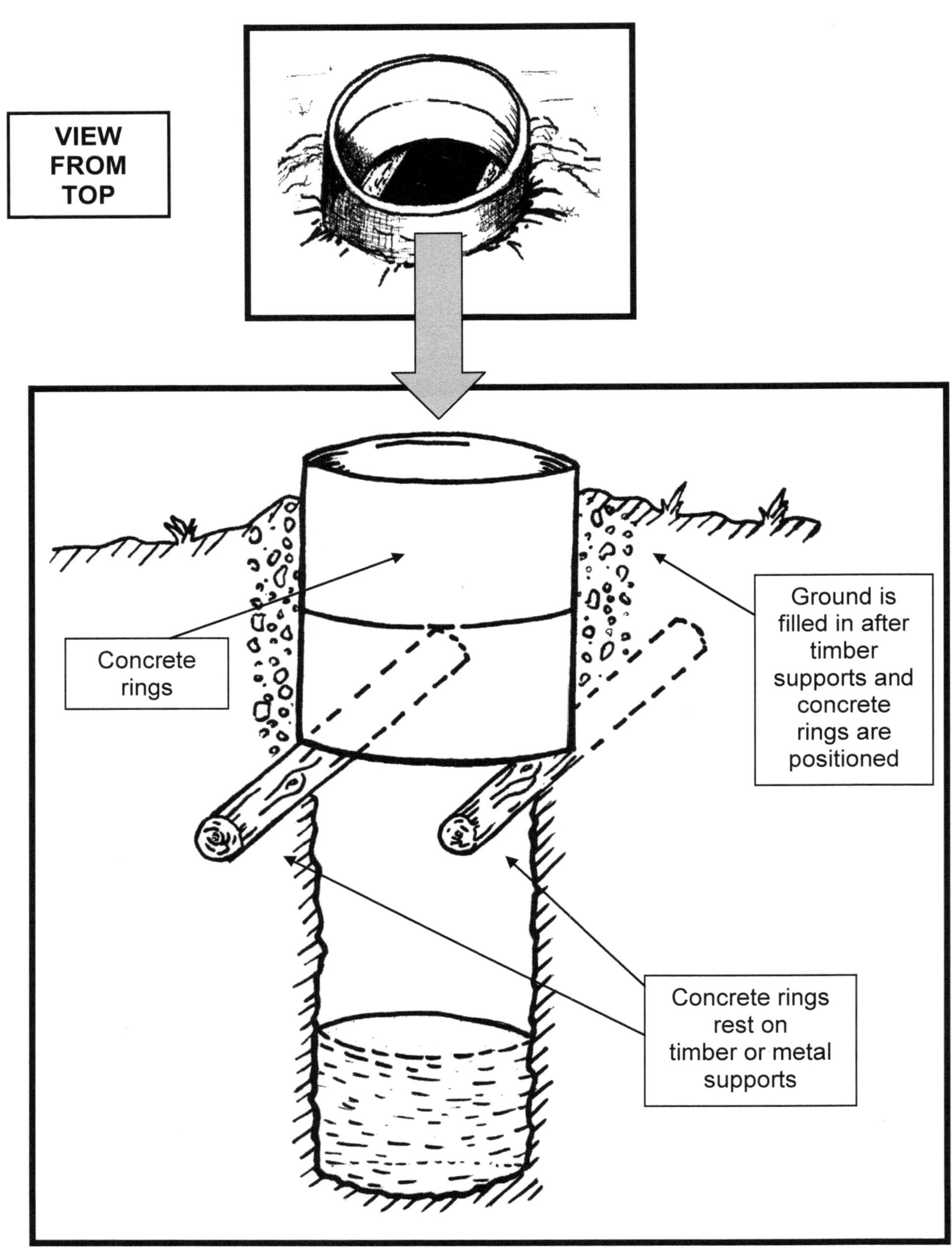

OPTIONS FOR LINING SHAFTS

Poster W. 34

FULLY LINED WITH CONCRETE RINGS

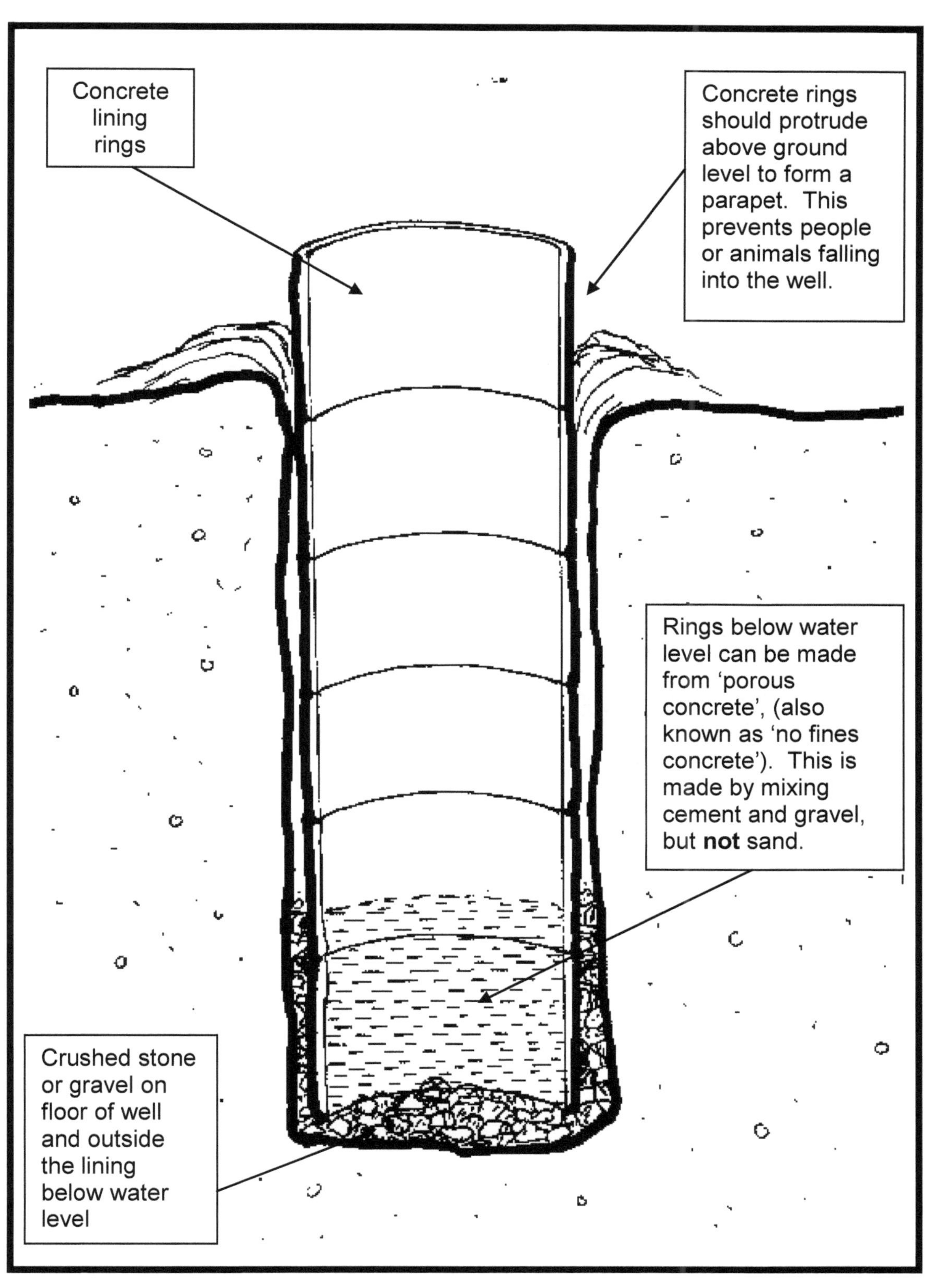

Poster W. 35
PARTIAL WOOD LINING

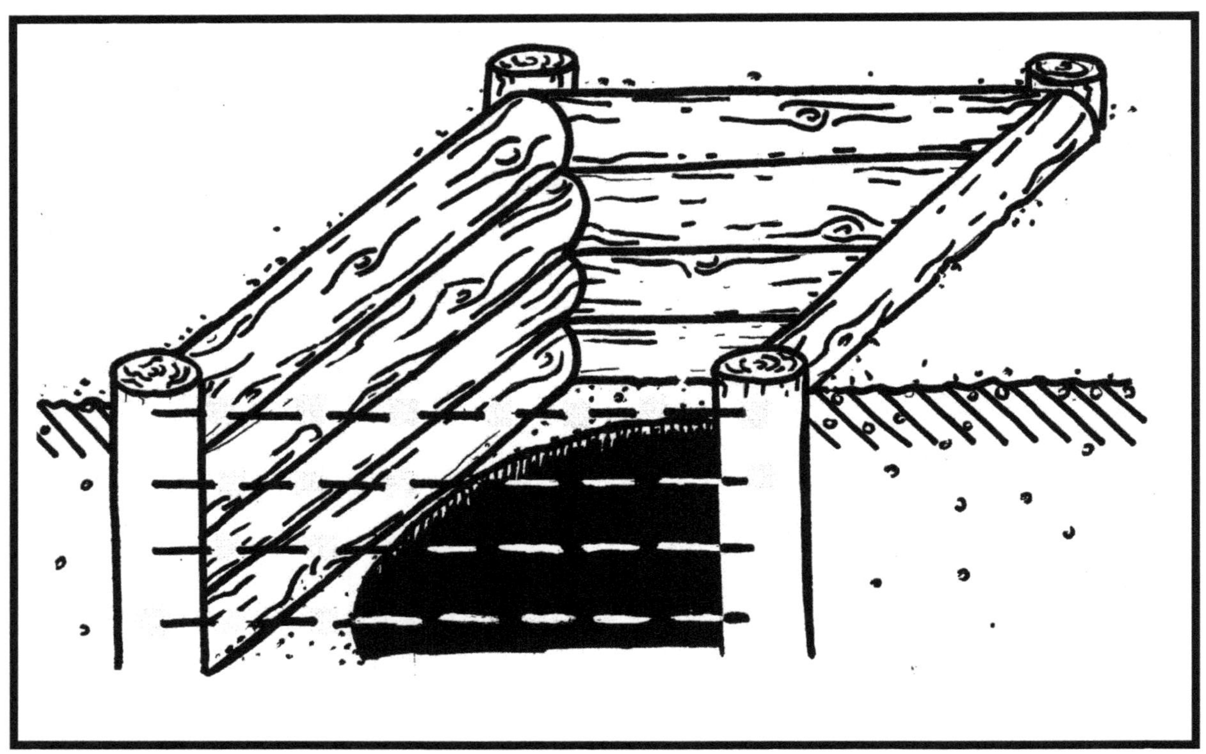

Poster W. 36

BASKET-WEAVE LINING FOR PITS

Woven lining can be made from twigs, grass, canes or branches interwoven around supports.

MINIMUM OPTIONS FOR TOP OF WATER SOURCES, STEP BY STEP IMPROVEMENT

Poster W. 37

UNPROTECTED WELL OR SCOOPHOLE

MINIMUM OPTIONS FOR TOP OF WATER SOURCES

Poster W.38

MOUTH OF SHAFT BUILT UP (MOUND) TO STOP FLOW INTO SOURCE

MINIMUM OPTIONS FOR TOP OF WATER SOURCES

Poster W.39

SOURCE CLOSED WITH OIL DRUM AND LID

MINIMUM OPTIONS FOR TOP OF WATER SOURCES

Poster W.40

TOP STRENGTHENED WITH BRICK TOP WALL AND SMALL APRON

MINIMUM OPTIONS FOR TOP OF WATER SOURCES

Poster W.41

SOURCE WITH APRON AND DRAINAGE (SOAK-AWAY)

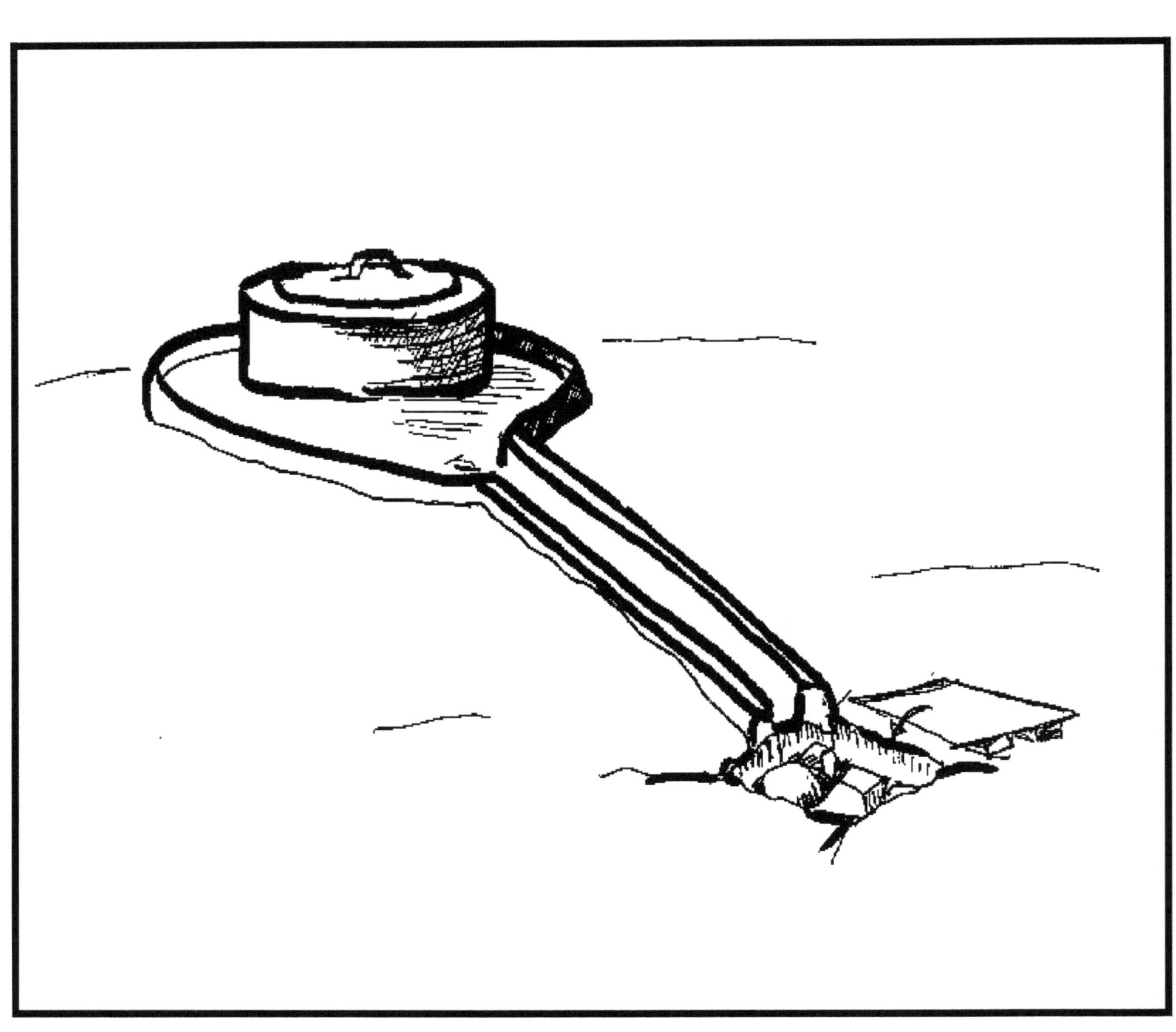

MINIMUM OPTIONS FOR TOP OF WATER SOURCES

Poster W.42

SOURCE WITH APRON, DRAINAGE AND GARDEN

MINIMUM OPTIONS FOR TOP OF WATER SOURCES

Poster W.43

POLE TO HANG BUCKET AND ROPE

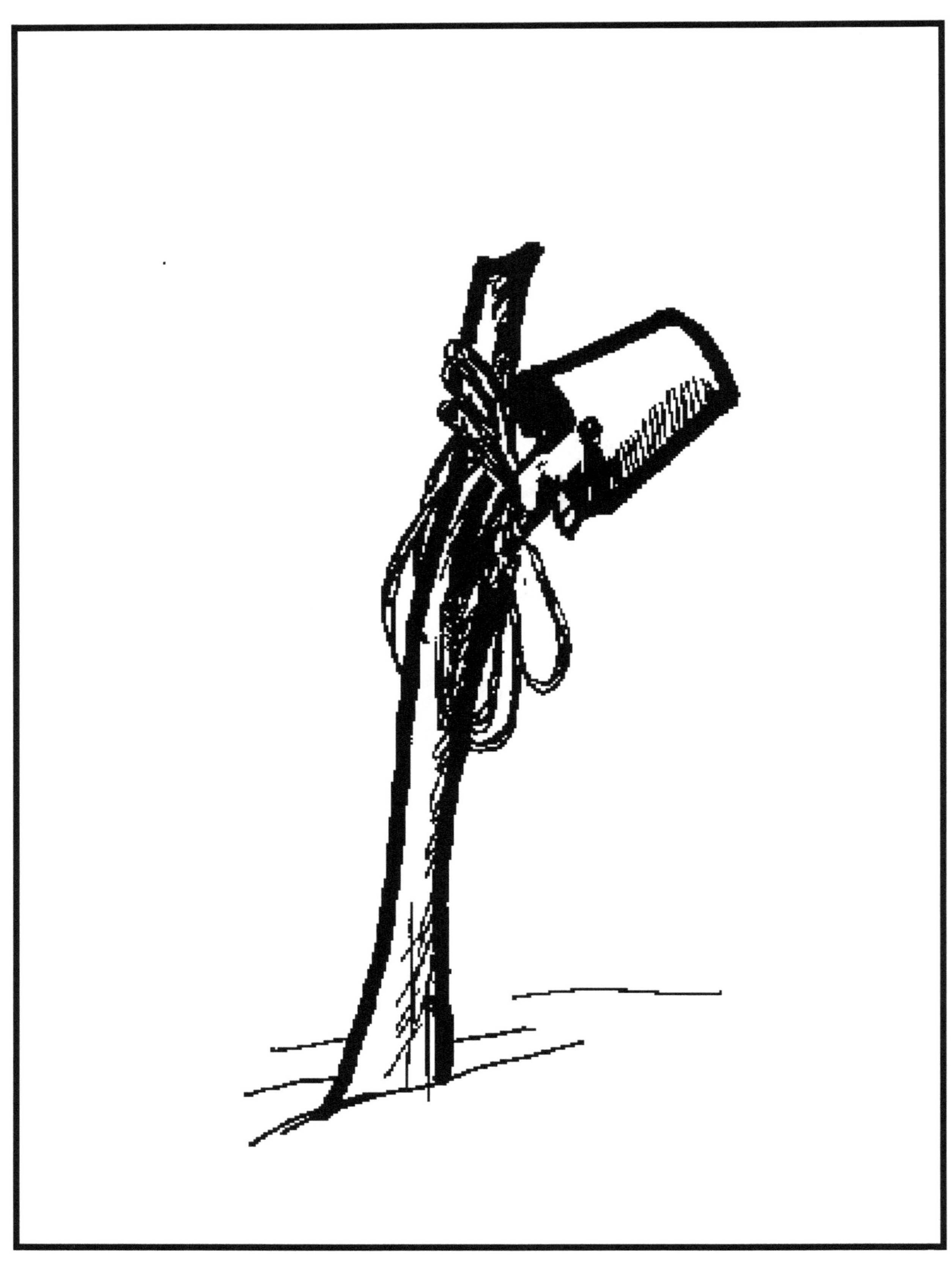

MINIMUM OPTIONS FOR TOP OF WATER SOURCES

Poster W.44

LOCALLY MADE WINDLASS

MINIMUM OPTIONS FOR TOP OF WATER SOURCES

Poster W.45

HOW A PULLEY WORKS

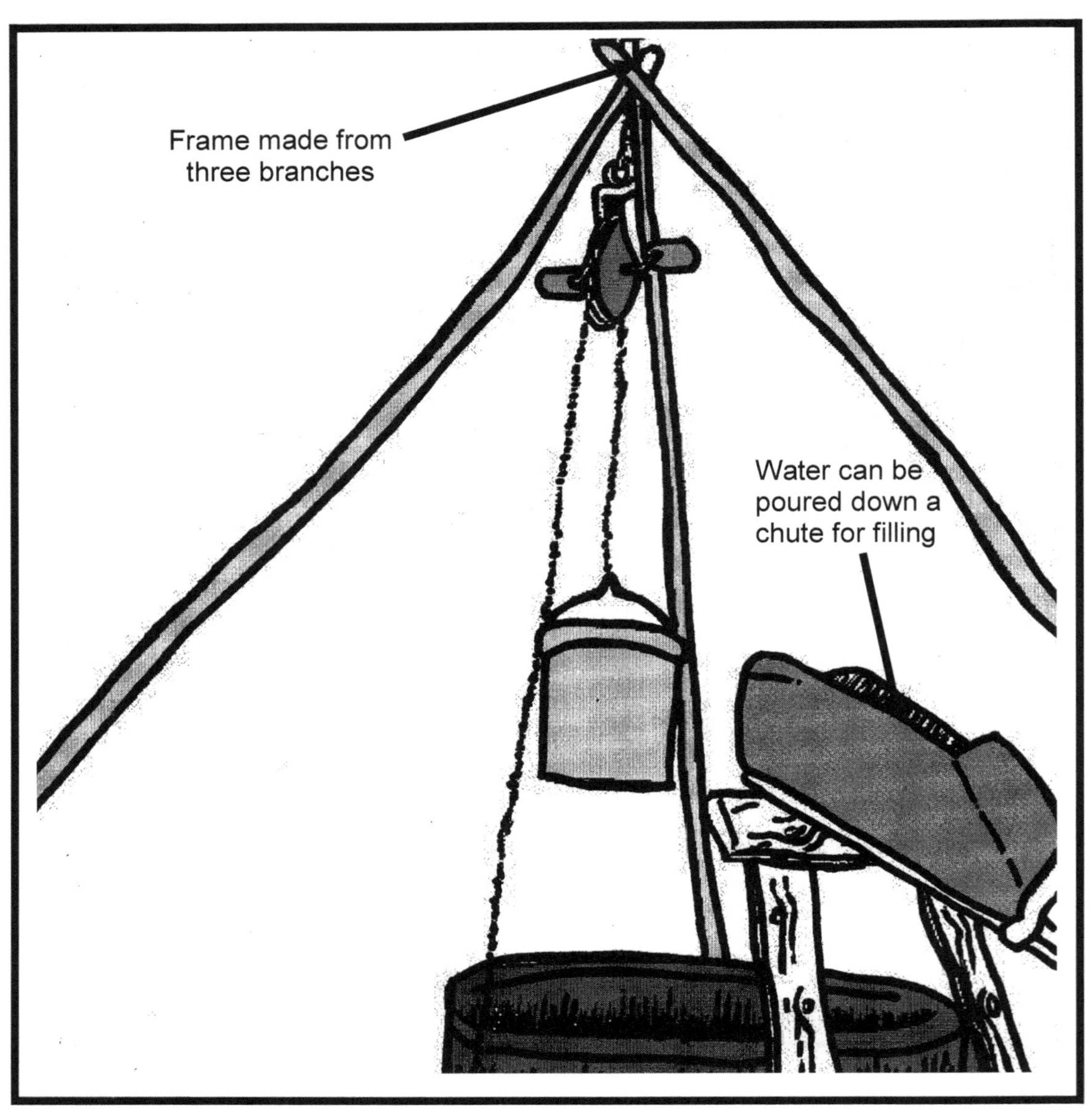

MINIMUM OPTIONS FOR TOP OF WATER SOURCES

Poster W.46
USING A PULLEY TO DRAW WATER FROM A WELL (1)

MINIMUM OPTIONS FOR TOP OF WATER SOURCES

Poster W.47

USING A PULLEY TO DRAW WATER FROM A WELL (2)

www.ingramcontent.com/pod-product-compliance
Ingram Content Group UK Ltd.
Pitfield, Milton Keynes, MK11 3LW, UK
UKHW050825070726
13597UKWH00021B/164